Climate Change and Sustainable Development

Published by
ACADEMIC FOUNDATION
in association with

Institute of Public Enterprise

Climate Change and Sustainable Development

Global Perspective

Editors

R.K. Mishra

P.S. Janaki Krishna

Ch. Lakshmi Kumari

ACADEMIC FOUNDATION

NEW DELHI

www.academicfoundation.org

First published in 2017
by

ACADEMIC FOUNDATION
4772-73 / 23 Bharat Ram Road, (23 Ansari Road),
Darya Ganj, New Delhi - 110 002 (India).
Phones : 23245001 / 02.
Fax : +91-11-23245005.
E-mail : booksaf@gmail.com
www.academicfoundation.org

in association with
Institute of Public Enterprise, Hyderabad.

Disclaimer:
The findings/views/opinions expressed in this book are solely those of the authors and do not necessarily reflect the views of the publisher.

Cataloging in Publication Data--DK
Courtesy: D.K. Agencies (P) Ltd. <docinfo@dkagencies.com>

Climate change and sustainable development : global perspective / editors, R.K. Mishra, P.S. Janaki Krishna, Ch. Lakshmi Kumari.
pages cm
ISBN 9789332704176

1. Climatic changes--India--Congresses. 2. Climatic changes--Congresses. 3. Sustainable development--India--Congresses. 4. Sustainable development--Congresses. 5. Natural resources--India--Management--Congresses. 6. Natural resources--Management--Congresses. 7. Sustainable agriculture--India--Congresses. 8. Sustainable agriculture--Congresses. 9. Renewable energy sources--India--Congresses. 10. Renewable energy sources--Congresses. I. Mishra, R. K., 1948- editor. II. Janaki Krishna, P. S., editor. III. Lakshmi Kumari, Ch., editor. IV. Institute of Public Enterprise (Hyderabad, India), organizer, publisher.

HC440.E5C55 2017 DDC 338.9270954 23

Typeset, printed and bound by The Book Mint, New Delhi.
www.thebookmint.in

CONTENTS

Section 2
Natural Resource Management for Sustainable Development

Section 3
Sustainable Agriculture

List of Tables and Figures

TABLES

FIGURES

About the Editors/Contributors

Editors:

Ram Kumar Mishra, Director, Institute of Public Enterprise, Hyderabad has research and teaching stints at London Business School; Maison Des Sciences De L' Hommes, Paris; University of Bradford, United Kingdom; UN Center on Promotional Enterprises, Ljubljana, as well as University of Ljubljana, Slovenia. He is a Member of the UN Task Force on International Taskforce on Standards of Excellence in Public Administration and Education, and is on the Steering Committee of several national and international bodies. Prof Mishra has also been a researcher for Planning Commission, Finance Commission, Prime Minister's Economic Advisory Council, State Bureaus of Public Enterprises, financial institutions and banks, and has to his credit several books and international papers and reports.

P.S. Janaki Krishna is an Associate Professor and Coordinator, Centre for Sustainable Development at Institute of Public Enterprise, Hyderabad. She has vast experience in research, teaching and technology transfer, and is experienced in dealing with various stakeholder organizations; she is also involved in technology monitoring and evaluation. She is one of the regular writers for an international news report "Information Systems for Biotechnology (ISB) News Report", being published by Virginia Polytech and State University, USA, and a Member of 'e-agriculture community' of FAO. Dr Krishna has authored books and several papers in national and international journals.

Ch. Lakshmi Kumari is an Assistant Professor in Economics at Institute of Public Enterprise, Hyderabad. She has been associated with the institution in various activities viz., teaching, training, research and consultancy assignments. Dr Kumari has worked for several projects at national and international level, and has published many reports. Her areas of interest include: Economics, privatisation strategies, restructuring and turnaround management issues, financial performance analysis, venture capital, environmental impact analysis, manpower rationalisation issues and recruitment process of manpower in companies.

Contributors:

D. Anil, Acharya N.G. Ranga Agriculture University, Rajendranagar, Hyderabad.

Dakuri Arjun, University College of Technology, Osmania University, Hyderabad.

V. Bhasker, University College of Technology, Osmania University, Hyderabad.

G.R. Chary, Principal Scientist (Agron.), Central Research Institute for Dryland Agriculture, Santoshnagar, Hyderabad.

Dakuri Chiranjeevi, Malla Reddy College of Engineering for Women, Hyderabad.

K.B. Eswari, Department of Genetics and Plant Breeding, College of Agriculture, Acharya N.G. Ranga Agriculture University, Hyderabad.

Omanjana Goswamia, Department of Applied Chemistry, Birla Institute of Technology, Ranchi, Jharkhand.

V. Hemalatha, Department of Genetics and Plant Breeding, College of Agriculture, Acharya N.G. Ranga Agriculture University, Hyderabad.

Usha Jha, Department of Applied Chemistry, Birla Institute of Technology, Ranchi, Jharkhand.

M.D. Mujeeb Khan, Principal Scientist and Head (Retired) IWM Unit, RARS, Lam Farm, Guntur.

A. Lalitha, Scientist, Acharya N.G. Ranga Agricultural University (ANGRAU), Hyderabad.

Prajnaparamita Mishra, Assistant Professor, School of Economics, University of Hyderabad, Hyderabad.

Sumit Mishra, Department of Applied Chemistry, Birla Institute of Technology, Ranchi, Jharkhand.

K. Mosha, Coordinator, DAATTC. Visakhapatnam, A.P.

J. Narender, Acharya N.G. Ranga Agriculture University, Rajendranagar, Hyderabad.

R. Naresha, Acharya N.G. Ranga Agriculture University, Rajendranagar, Hyderabad.

Josily Samuel, Scientist, Central Research Institute for Dryland Agriculture, Santoshnagar, Hyderabad.

B.N. Sandeep Nayak, P.G. Student, Department of Agronomy, Agricultural College, Bapatla, Andhra Pradesh.

M. Osman, Principal Scientist (Agron.), Central Research Institute for Dryland Agriculture, Santoshnagar, Hyderabad.

G. Padmaja, Acharya N.G. Ranga Agriculture University, Rajendranagar, Hyderabad.

J.V.N.S. Prasad, Principal Scientist (Agron.), Central Research Institute for Dryland Agriculture, Santoshnagar, Hyderabad.

K.G.R. Prathyusha, Research Associate, Central Research Institute for Dryland Agriculture, Santoshnagar, Hyderabad.

Puspanjali, Scientist, Central Research Institute for Dryland Agriculture, Santoshnagar, Hyderabad.

P. Steven Raj, Centre for Economic and Social Studies, N.O. Campus, Begumpet, Hyderabad.

A. Rakesh, Department of Genetics and Plant Breeding, College of Agriculture, Acharya N.G. Ranga Agriculture University, Hyderabad.

T. Ramanjali, Acharya N.G. Ranga Agriculture University, Rajendranagar, Hyderabad.

V.G. Ranganath, Assistant Professor, Faculty of Law, ICFAI Foundation for Higher Education (Deemed to be University), Hyderabad.

P. Prasuna Rani, Associate Professor, Department of Soil Science and Agriculture Chemistry, Agricultural College, Bapatla, Andhra Pradesh.

K.V. Rao, Principal Scientist (SWCE), Central Research Institute for Dryland Agriculture, Santoshnagar, Hyderabad.

P. Padmanabha Rao, Associate Professor, Centre for Economic and Social Studies, Nizamiah Observatory Campus, Hyderabad.

V.U.M. Rao, PC (AICRPAM), Central Research Institute for Dryland Agriculture, Santoshnagar, Hyderabad.

B. Suresh Reddy, Centre for Economic and Social Studies, N.O. Campus , Begumpet, Hyderabad.

M. Gopinath Reddy, Professor, Centre for Economic and Social Studies, N.O. Campus, Begumpet, Hyderabad.

R. Rejani, Senior Scientist (SWCE), Central Research Institute for Dryland Agriculture, Santoshnagar, Hyderabad.

K.R. Sastry, Formerly Professor & Head, Dr. MCR HRD Institute of Telangana; now Consultant, Disaster Management.

Gautam Sen, Department of Applied Chemistry, Birla Institute of Technology, Ranchi, Jharkhand.

Shalini Sharma, Chevening Rolls-Royce Science, Innovation, Policy and Leadership Fellow–2014, University of Oxford, UK, and Fulbright-Nehru Environment Leadership Fellow–2012, Earth Institute-Columbia University, USA.

V. Swarnalatha, Department of Genetics and Plant Breeding, College of Agriculture, Acharya N.G. Ranga Agriculture University, Hyderabad.

Ch. Srilatha Vani, Scientist, Acharya N.G. Ranga Agricultural University (ANGRAU), Hyderabad.

Acknowledgements

We thank all the paper contributors for their interest and support in bringing out these proceedings in the form of book. We thank Dr B. Venkatshwarlu, Former Director, Central Research Institute for Dryland Agriculture (CRIDA), Prof Ch. Sailu, Principal, University College of Technology, Osmania University and Prof V.V. Basava Rao, Coordinator–Technical Education Quality Improvement Program (TEQIP–II), University College of Technology, Osmania University for their constant support and guidance in bringing out this book. We thank the paper contributors for their enormous patience in waiting for this publication to see the light of the day.

—R.K. Mishra, P.S. Janaki Krishna and Ch. Lakshmi Kumari

Preface

We are happy to place this book *Climate Change and Sustainable Development: Global Perspective* which includes the papers presented and selected for publication at a two-day international conference jointly organised by Institute of Public Enterprise (IPE), Hyderabad in collaboration with Technical Education Quality Improvement Programme, University College of Technology (UCT), Osmania University (OU), Hyderabad, Central Research Institute for Dryland Agriculture (CRIDA), Hyderabad and 500 Eco, San Jose, USA.

The United Nations Rio+20 Summit and the commitments of governments to create universal SDGs (sustainable development goals) have generated new and fresh debate on climate change and sustainable development, and issues and challenges on how to best pursue environmental sustainability with social equity. Sustainable development and climate change are themes which operate on local, regional, national and global scales and require multidisciplinary, cross-sectoral research and cooperation. Environmentally, socially and economically sustainable development as well as successful climate change adaptation and mitigation can be achieved only by encouraging knowledge sharing and cooperation between various sectors and decision-makers. In India, the role of government, private and NGOs will be very significant for the success of such goals, and therefore, it is important to mainstream them into the planning process. It is also necessary to have clear means of implementation for the SDGs, which can be the building blocks including financing, governance, and availability of technology. India needs to strengthen capacity of elected representatives and government officials at state, district and local levels, and focus on strategic government development programmes and schemes which are geared towards achievement of SDGs and national development goals. Thus, the Conference was designed to address some of the key issues in climate change and sustainable development The blind reviewed papers presented in the Conference and selected for publication are presented in this book.

The papers covered wide array of areas in climate change mitigation and adaptation from carbon emission trading to advances in corporate social responsibility and environmental regulations. The papers discussed on emerging insights from constraints in practicing organic farming and impact of land use on urban environment for sustainable development and concepts of green buildings mainstreaming disaster risk reduction initiatives. As climate change and sustainable development pose a serious threat to human survival, food security, natural resources, human health, economic activity and physical infrastructure, the following issues were presented and discussed in the book.

As the concept of ETS (emission trading scheme) is crucial in regulating global emissions the book outlined various ETSs and examined their operational efficiency in various countries. The growth of organic farming is found to be losing ground in spite of the efforts being made by the well wishers of people and environment, government and agricultural universities. A few conventional farmers are entering organic farming but equally existing organic farmers lost their hope in organic farming and started practicing both organic farming and conventional farming practices. Hence, a study has been presented in the book with selected sample organic farmers to know about the constraints faced by the farmers. In the book salient aspects relating to human development; human activities for settled agriculture and its impact to environment; impact of modern agriculture on climate change; city development and urbanization; industrial revolution were presented. The forcing factors that cause climate change viz., usage of heavy metals, CFCs (Chloro Floro Carbons), oxides of nitrogen and sulphur, depletion of ozone layer, radioactive substances and their implications were also presented and it also touched upon the effect of climate change on human health and finally reiterated the need for strengthening sustainable governance.

The structure of textile industry, its impact on environment, its pollution control strategy, and response of Ministry of Textiles are also dealt in the book. Effective pollution control strategies such as: reduction in waste water, counter current washing operations, optimum use of process chemicals, substitution of toxic chemicals by eco-friendly chemicals, good housekeeping, recovery of chemicals such as caustic soda, dyes and grease, replacement of high BOD chemicals by low BOD ones, use of low liquor ratio dyeing equipment, improved production processes to eliminate and reduce waste as much as possible were suggested as a solution to the environmental problem. In the book it was discussed that precautionary principle plays a significant role in determining whether developmental process is sustainable or not. Precautionary principle underlies sustainable development which explains that the developmental activity must be stopped and prevented if it causes serious and irreversible environmental damage. The emergence of precautionary principle marks a shift in the international environmental jurisprudence—a shift from assimilative capacity principle to precautionary principle. As special and temporal variation of rainfall is important the study in low, medium and high rainfall districts was also presented in the book. A robust high performance buildings movement to rethink the built environment is rapidly emerging and affecting the design, construction and operation of new buildings; changing the renovation process for existing buildings; and reshaping cities and communities. Thus, the concept of green buildings was presented as paradigm shift in towards sustainable development.

The papers on sustainable agriculture mainly discussed the impact of climate change on agriculture and also the resource recovery mechanism of wetlands. The book also focused on corporate social responsibility towards sustainable development wherein initiatives taken up by the sustainable development Division of Hindustan Unilever Limited (HUL) were presented. The study found that CSR in India tends to focus on what is done with profits after they are made. Hence, much of the Indian practice of CSR is an important component of sustainability or responsible business, which is a larger idea, a fact that is evident from various sustainability frameworks.

Initiatives of CSOs in mainstreaming disaster risk reduction clarified that activating Early Warning System (EWS) Network and its close monitoring mechanisms for integrating the scientific, technological and administrative agencies for effective disaster reduction was essential. It also reiterated that development models have to be participatory and partnerships need to be built between various stakeholders. More importantly the book also dealt in detail the plethora of environmental regulations for sustainable development. The book thus covers diverse areas of interest in climate change for sustainable development.

Some speakers provided only extended abstracts of their presentations hence we were constrained to include their full papers. Nevertheless, we assure that the issues raised and covered in this book are current as they have been all these years and they continue to be so for some time to come. All the opinions expressed in the book strictly belong to the respective authors who contributed the papers. Neither the editors nor IPE holds any responsibility.

It is our hope that this book will provide useful insights to the readers with regard to the status on climate change and sustainable development and provides a direction to all the stakeholders in meeting the needs to deal with the twin issues of climate change and sustainable development.

—R.K. Mishra, P.S. Janaki Krishna and Ch. Lakshmi Kumari

Section 1

R&D in Climate Change and Sustainable Development

1

R.K. MISHRA, P.S. JANAKI KRISHNA AND
CH. LAKSHMI KUMARI

Emission Trading Practices

A Global Scenario

Introduction

The concept of Emission Trading Scheme (ETS) is the outcome of Kyoto Protocol of 1997 that gave rise to the importance on the use of ETS in order to reach the goals set by the protocol. ETS is a mechanism to regulate global emissions by government imposing caps. (Measuring units are fixed for these caps viz., tonnes of carbon dioxide [CO_2] emission.) Government acts as a regulator to set limit or cap on the amount of a pollutant that may be emitted. Therefore various countries across the globe have been adopting ETS as one of the strategies for mitigating climate change. The present paper outlines various ETS and examine their operation and efficiency in various countries. It further provides an overview of the global experiences of ETS.

There are two main types of ETS viz., cap & trade schemes, baseline & credit schemes that can effectively regulate access to restricted resources. The trading mechanism results in a market-based signal that determines the price for emission. Under the market-based mechanism, if the cost of avoiding emissions are less than what the participants receive if they sell allowances or credits, participants will avoid emissions and then sell allowances (in a cap & trade scheme) or credits (in a baseline & credit scheme). On the other hand, if the costs of avoiding emissions exceed what participants have to pay to buy the equivalent amount of allowances or credits, participants will emit and will buy allowances or credits to pay for those emissions.

The major difference between cap & trade schemes and baseline & credit schemes is that cap & trade schemes implement the cap on emissions by issuing allowances to emit up to the cap, while baseline & credit schemes implement the cap on emissions by assigning individual baselines to participants up to the cap. As an increasing number of countries, regions, cities and states implement emissions trading policies to limit CO_2 emission, many may turn to the experience of the European Union's ETSs, the largest GHG (greenhouse gas) emissions trading system in operation

These schemes are further divided into: a) statutory schemes, and b) non-statutory schemes. Statutory schemes are mandatory and the non-statutory are voluntary. For instance, European Union GHG Emission Trading Scheme and New South Wales GHG Reduction Scheme (GGAS) are mandatory where as CCX and CDM are voluntary. Table 1.1 gives an example for four possible combinations.

Table 1.1

Combination of Emission Trading Schemes

	Cap & Trade Schemes	*Baseline & Credit Schemes*
Statutory schemes (mandatory)	European Union Greenhouse Gas Emissions Trading Scheme (EU ETS)	New South Wales Greenhouse Gas Reduction Scheme (GGAS)
Non-statutory schemes (voluntary)	Chicago Climate Exchange (CCX)	Clean Development Mechanism (CDM)

Source: Compiled from literature.

Emission Trading Scheme: Global Scenario

ETS requires several other elements to function smoothly, especially in terms of administrative and monitoring capacity. In general, the construction of a carbon trading system involves certain main components: i) prudent GHG accounting and verification rules; ii) trading infrastructure, such as registries and exchanges; and iii) an accountability system in case of non-compliance. Each of these components is indispensable, and together they require not only creditable carbon emission measurement and statistics, but also a fair allocation mechanism, free market conditions and reliable oversight, as well as strict monitoring. Any one of those aspects has the potential for misuse and rent-seeking, thus corrupting the system as a whole and rendering it ineffective.

The performance of carbon emission trading markets is normally measured by the resulted emission reduction (effectiveness), the emission reduction cost (cost-effectiveness or efficiency), innovation and investment in clean technology, and any resulting carbon leakage (Lu 2011). Since the emergence of the EU ETS in 2005, the world has seen a rapid growth in carbon markets. In 2011, the global carbon market reached US$142 billion. Despite the recent 'bad patch' due to a range of reasons from Europe's economic downturn to the uncertainty of the future of CDM, which remains uncertain until a new climate agreement has been reached, many believe that carbon markets remain one of the most trusted and attractive instruments we have to combat climate change (Fankhauser 2011: 35).

Today number of countries are integrating cap-and-trade systems into their national climate policies. The EU ETS, is the frontrunner in this development. Further large number of emissions trading systems both at National and State level are emerging around the world. These initiatives have been launched at the state level in US. The Regional Greenhouse Gas Initiative (RGGI) on the East Coast, the Western Climate Initiative (WCI) between states on the West Coast and other US states and Canadian provinces, and the Midwestern Greenhouse Gas Accord in the central part of the US. In addition, several legislative proposals for a federal system are currently under discussion in the US Congress. In Australia, detailed provisions for the scheme was started in 2010. New Zealand and Japan are also adopting the same. The present ETSs vary significantly in their size, design characteristics and geographical scope, sectoral coverage and the amount of credit limit.

The European Commission has been trying to make ETS as nucleus for creating a global carbon market. It has introduced OECD wide carbon market in 2015, and the priority was to establish a transatlantic link between the EU ETS and a federal US scheme. A combined EU-US market would cover the larger part of OECD emissions, and could thus constitute the backbone for the future international emissions trading regime. If the EU and US find common ground on key design elements, this would probably exert significant influence on the other. The other countries like Switzerland, Australia, South Korea, New Zealand and Kazakhstan are part of ETS.

In 2008, Swiss ETS came into being for complying national CO_2 levy on heating, industrial process, and transport fuels. In the year 2013, the ETS has become mandatory. New Zealand also started ETS to cover forestry and further expanded to cover energy, fishing industrial processes and fossil-fuels. Australia started implementing CPM (carbon price mechanism) as ETS in the year 2012. Under the scheme, 500 companies will pay a tax of A$23 per tonne of carbon, rising by around 5 per cent every year. The price is fixed in the initial years. From 2015 onward, the permits price was under a cap-and-trade system.

Table 1.2

Major Carbon Trading Schemes

Kyoto Protocol	Established in 2005. The target is to reduce emissions by 5 per cent below 1990 levels in 2008-2012, collectively, for the 37 countries with mandatory targets. Countries can reduce emissions at home, buy permits from one another, or buy carbon offsets from projects in developing countries under the CDM.
European Union Emissions Trading Scheme	Established in 2005, mandatory for all 27 EU members, plus Iceland, Liechtenstein and Norway, covering about half of total EU carbon emissions. The target is to reduce emissions by 21 per cent below 2005 levels by 2020. Under the scheme, member states allocate a quota of emission allowances to 11,000 industrial installations. Companies get most permits free during the second phase based on historical emissions, but many electricity generators have to pay for all these from 2013 (third phase) . More than 3,000 airline operators have joined the scheme in 2012.*
New Zealand emissions trading scheme	Launched July 1, 2010. Mandatory, with the target to cut greenhouse gas emissions between 10 and 20 per cent by 2020 on 1990 levels. Under the scheme, emissions units are allocated based on an average of production across each industry. Sectors include forestry, electricity, industrial process emissions and transport, waste (start 2013), and agriculture (start 2015). From 1 July 2010, to 1 January 2013, emitters have the option of paying a fixed price of NZ$25 per tonne of carbon, and will only have to surrender 1 unit for every 2 units of emissions.

Northeast U.S. Regional Greenhouse Gas Initiative	Launched January 2009, covering carbon from power plants in 10 states in the US Northeast. The target is to reduce emissions by 10 per cent below 2009 levels by 2018. Allows offsets from five different types of clean energy projects, including capturing methane from landfills and livestock manure, but only if a US$7 per tonne price trigger is hit
Japan: Tokyo metropolitan trading scheme	Launched in April 2010, covering around 1,400 top emitters in the metropolitan area. Japan aims to cut emissions by 25 per cent by 2020 from 1990 levels. Under the scheme, Tokyo sets emission limits for large factories and offices, which can be met by using technology such as solar panels and advanced fuel-saving devices.
Australia: Clean Energy Bill	Adopted by the Australian Parliament, covers emissions from all sources except agriculture and land use, or the combustion of biomass, bio-fuels and bio-gas. The national target is to cut emissions by 5 per cent below 2000 levels by 2020. Under the scheme, 500 companies will pay a tax of A$23 per tonne of carbon from July 2012, rising by around 5 per cent a year, and move to a market based trading scheme in 2015.
California climate change law (AB32)	Launched in 2013, with the first permit auction in November 2011. Covers emissions from power plants, manufacturing and transportation fuels. The target is to cut the state's emissions to 1990 levels by 2020. Most of credits will be allocated free in the early years, and emitters will be allowed to use offsets to fulfil up to 8 per cent of their compliance obligation
Western Climate Initiative	Launched in January 2013, covering California, Canada's British Columbia and Quebec, and possibly also Ontario. The target is to cut emissions by 15 per cent below 2005 levels by 2020. Under the scheme, emitters such as power plants will have to buy offsets to cover their emissions.
South Korea emissions trading scheme	Started in 2015 covering about 470 companies from all sectors that together produce about 60 per cent of the country's emissions. The South Korean government has set a 2020 emissions reduction target of 30 per cent below projected "business as usual" levels
India: Perform, Achieve and Trade system	Started in 2014 after a three-year rollout period. It is a mandatory energy efficiency trading scheme covering eight sectors responsible for 54 per cent of India's industrial energy consumption. India has pledged 20 to 25 per cent reduction in emissions intensity from 2005 levels by 2020. Under the scheme, annual efficiency targets will be allocated to firms. Tradable energy-saving permits will be issued depending on the amount of energy saved during a target year
China: Pilot carbon trading schemes	In November 2011, China approved pilot tests of carbon trading in seven provinces and cities—Beijing, Chongqing, Guangdong, Hunan, Shanghai, Shenzhen and Tianjin. Some of the pilot regions started trading as early as 2013/14.

Source: Authors' summary of Reuters factbox; **:Reuters (2011). "Carbon Trading Schemes Around the World". Available at *http://www.reuters.com/article/2011/07/11/us-carbon-schemes-idUSTRE76A2GJ20110711.*

In 2013 ETS was launched in the Republic of Kazakhstan. Manufacturing, energy, mining, metallurgy, chemicals, agriculture and transport industries were covered which emit more than 20,000 tonnes of CO_2 per year.

China today accounts for almost a quarter of global greenhouse gas emissions and about half of annual emissions growth (Steckel et al. 2011). Moreover, such trends are expected to continue. The International Energy Agency (IEA 2009) has estimated that about half the growth in global energy-related CO_2 emissions from now until 2030 will come from China. China's emission growth trajectory is therefore a critical factor in the world's chances of keeping global warming below 2°C. In its most recent global carbon emission trends report, the Joint Research Centre of the European Commission and the Netherlands Environmental Assessment Agency (Olivier et al. 2011) found that China's per capita CO_2 emissions had already reached 6.8 tonnes, higher than France's and on a par with Italy's. Table 1.2 gives a summary of the existing and major emerging carbon trading schemes in the world.

The key features of ETS in various countries and their operational efficiency is analysed hereunder.

Key Features of Australian Emissions Trading Scheme

In December 2008 the Australian government published a White Paper regarding the introduction of an ETS on Australia. The proposed 'Carbon Pollution Reductions Scheme' has very broad coverage which includes all GHGs and 75 per cent of Australia's emissions. Forestry is covered by a voluntary opt-in approach from the beginning. Around 1000 entities above 25 kt CO_2-eq/year are directly liable for their emissions, and smaller sources of combustion emissions through 'upstream' permit liability on fuel suppliers. In this way, transport and household emissions are also covered by the price signal created in the permit market, although some rebates are provided in early years in the transport sector. Permits are to be auctioned except for free allocations to Energy Intensive Trade Exposed Industries (EITEI) such as aluminium, steel and liquefied natural gas and a defined amount to be granted to coal-fired electricity generators. The free allocation to EITEI is amounting to 90 or 60 per cent of historic benchmark emissions in the sector and includes new entrants. These kind of industries account for a much larger share in the economy and aggregate emissions in Australia than in most other developed countries. These industries further challenges for emissions trading design to address the threat of carbon leakage while avoiding an adverse political economy of discretionary handouts by government to industry.

The decision to give an uncapped amount of free permits to trade-exposed industries simply on the basis of emissions intensity including new entrants has been one of the main points of criticism by domestic countries about the scheme.

Australia's unconditional national emissions target commitment is to be a 5 per cent reduction by 2020. Conditional on commitments undertaken by other countries, the proposed target range extends to a maximum national commitment of a 15 per cent reduction. A price cap was applied in the period 2010-2015, starting at A$40/t and rising at 5 per cent per year plus adjustment for inflation. The government could sell additional permits into the market at this fixed price. If and when in place, the price cap would loosen the Australian Scheme cap, and the government would have to buy additional Kyoto units from the international market to meet its international obligations possibly at a higher price than the price cap, creating potential budgetary exposure. International Kyoto credits can be purchased in an unlimited quantity, trading by scheme participants is to be restricted however to purchases of non-forestry CDM, JI and removal units, subject to future review. It is to be expected that the bulk of permit purchases by Australian emitters will be from the CDM. Permit sales from the Australian system into overseas systems are specifically excluded as transferring Australian permits to other countries would increase the Australian permit price resulting in relatively more abatement occurring in Australia than would otherwise be the case. These rules cap the permit price in Australia at the international CDM price, the price cap, or the domestic price if no international purchases are necessary, whichever is lowest.

Key Features of the New Zealand Scheme

The New Zealand emissions trading scheme is the first to place a cap on emissions from deforestation. Under the proposed scheme, NZUs need not submit for GHGs released as a result of harvesting and regrowth cycles. In addition, landowners whose deforest is less than two hectares need not report (or submit NZU's) and owners of less than 50 hectares of forestland can apply for an exemption from the system. Methodologies for determining emissions from deforestation will be provided by the administering agency. Currently New Zealand has no price cap in place as it expected a large supply of international credits and therefore a moderate price of these credits.

The Indian Perspective of Emission Trading Scheme

India is currently the third largest polluting country in the world, emitting 2407 million CO_2 as of 2015. It is expected to contribute 6 per cent of global GHG emissions by 2020. India ratified the Kyoto Protocol in 2002; however, as a developing country, it was not required to submit any obligatory reduction commitments. India did submit, in 2009, a voluntary target for reducing the emissions intensity of its GDP by 20-25 per cent relative to 2005 levels by 2020. The basis of India's climate policy framework is its 2008 National Action Plan on Climate Change (NAPCC), which has specified eight national objectives for 2017 that centre on improving energy efficiency, solar technology, sustainable habitats, water, Himalayan ecosystems, "green India", agriculture, and strategic knowledge. India's 'Perform Achieve and Trade' (PAT) initiative, which resembles an ETS, is currently at its end in its first phase (2012-2015), which is considered a test phase. What distinguishes India's PAT from traditional cap-and-trade systems is that cap-and-trade usually entails an absolute cap, whereas PAT specifies energy targets that are intensity based. India also has a Renewable Energy Certificate (REC) trading system, which is a non-ETS, market-based mechanism to fight climate change. While pilot ETS programmes which focus on the abatement of particulates, but not on CO_2 are being launched in the three states Tamil Nadu, Gujarat, and Maharashtra. Government of India has made absolute emission reduction as mandatory on the ground that climate change is a problem caused by developed countries.

According to Upadhyaya (2010), if such a mentality were to continue to dominate, "it would not be possible to achieve consensus for a cap-and-trade system in the near future." While India has not yet established a carbon market or carbon pricing policy, it does have in place a 'Perform Achieve and Trade' (PAT), which promotes energy intensity improvement, and a Renewable Energy Credit (REC) trading system. The interrelationships between the two programs are currently being deliberated, and eventually credits from either system may become fungible. In the near future, India has mandated the implementation of pilot emissions trading systems, which will aim to reduce emissions of particulates, in three states: Tamil Nadu, Gujarat, and Maharashtra. In October 2013, the Pollution Control Board for these regions released guidelines for stationary sources to utilise Continuous Emissions Monitoring Systems (CEMS) to measure emissions.

Perform Achieve and Trade

PAT has been likened to a 'Tradable White Certificate' (TWC) system. TWC systems are designed to achieve energy intensity targets at lowest cost through trade in energy savings certificates. This is in contrast to an ETS, where absolute emissions reductions are achieved at lowest cost through trade in emissions reductions certificates. While PAT was launched as a program stemming from 2008 National Action Plan on Climate Change (NAPCC), its structure flows from the Energy Conservation Act of 2001 (ECA-2001), which requires 15 energy-intensive sectors to implement energy efficiency measures. PAT covers eight of these sectors. ECA-2001 also details the process for purchasing Energy Saving Certificates (ESC) within the PAT system for compliance in order to avoid defaulting on obligations. For the PAT programme, ECA-2001 was amended to enable trading of ESC. PAT sets mandatory energy efficiency targets on 478 facilities that are either part of energy-intensive industries or members of the electricity sector, which together comprise about 60 per cent of India's 2007 GHG emissions.

Facilities covered by PAT are called "Designated Consumers" (DC) and the list of these facilities is published annually by Bureau of Energy Efficiency (BEE). PAT, energy aims to reduce emissions by 26 million tons of CO_2, as well as save 6.6 million tons of oil equivalent over its first commitment period (2012-2015) through energy efficiency measures. Covered facilities are generally obligated to improve energy efficiency by 1-2 per cent per year. Within PAT, energy efficiency targets are measured in terms of Specific Energy Consumption (SEC), for which baselines are determined for April 2007 to March 2010. The average SEC reduction target under PAT is 4.8 per cent per installation, and it expected that achieving this target will cost industry over USD 5.4 billion. Installations must achieve their plant specific targets within three-year compliance periods. An installation that fulfils and exceeds its SEC target will be able to sell ESC for the amount of its surplus energy improvements to installations that are unable to meet mandatory targets. Trading will occur via regulated exchanges; platforms for trading ESC have been designated in the two power exchanges Indian Energy Exchange (IEX) and Power Exchange India Limited (PXIL), and BEE has also set up a registry and exchanges for the trading of ESCerts. BEE hopes to enable cross-sectoral use of ESC. Companies that purchase ESCerts would do so in order to achieve compliance obligations and avoid non-compliance penalties. BEE issued guidelines and regulations in March 2012, and the issuance and trading of ESC began after April 2013. The Energy Security Act (2001) provides legal basis for the sale and purchase of ESC. To create market liquidity and price discovery before the market is launched, some ESC will be auctioned ex-ante, other ESCs will be allocated freely to companies, and individual facility targets will be set. Rules regarding banking have not yet been determined.

As shown in Table 1.3, the PAT covers facilities in nine sectors: thermal power, iron and steel, cement, fertilizers, textiles, aluminium, pulp and paper, Chlor-Alkali and railways. The railway sector has been included in the second phase. A facility is regulated as a DC (designated consumer) if it exceeds the sector-specific threshold for annual energy consumption.

Table 1.3

Sector Wise PAT Facilities Covered

Industry Sector	*Identified DCs*	*Annual Consumption Norm to DCs*
Aluminium	10	7,500
Cement	85	30,000
Chlor-Alkali	22	12,000
Fertilizers	29	30,000
Pulp & Paper	31	30,000
Thermal Power	144	30,000
Iron & Steel	67	30,000
Textiles	90	3,000
Railways	8*	N/A

Note: *: Number of probable DCs.

RDSO, Indian Railways: Energy Audit Manual for Indian Railways

The PAT system splits the facilities it regulates into categories based on a plant's energy efficiency potential. The system includes modern, state-of-the-art facilities as well as older, less efficient facilities. Due to a large diversity in plant designs, vintage of plants, scale of production and process technology, plants are grouped into sectoral bands with a broad target SEC range. This was done in recognition of the heterogeneity of plant types to avoid the closure of inefficient plants under the enforcement of a specific Sectoral Energy Consumption (SEC) target. However, SEC targets are calculated for each plant; more efficient plants have lower SEC targets to achieve and more inefficient plants higher SEC targets.

Pilot Emissions Trading System

India's two-year pilot ETS mechanism was unveiled from 1 February 2011, and three states—Gujarat, Tamil Nadu, Maharashtra have received government mandates to implement programmes. While the pilot ETS mechanism focuses on particulates, such as SO2, NOx, and Special Particulate Matter (SPM), which are detrimental to human health. The pilot ETS mechanism was launched by India's Ministry of Environment and Forests (MoEF) together with the country's Central Pollution Control Board (CPCB) and relevant

State Pollution Control Boards (SPCBs). According to the system's design, the SPCBs will determine which pollutants to include and set caps for industry facilities based on desired overall pollutant concentrations. State regulators will then distribute emissions permits to capped facilities, which have the option of either complying with their caps and selling extra permits, or buying from the market the amount of permits by which they exceed their caps. Experimenting with ETS is two-fold: i) it is a cost-effective method of emissions mitigation, and ii) it spurs innovation.

The purpose of this programme is to improve air quality and to incentivise facilities to do their part in enabling the states as a whole to meet the National Ambient Air Quality Standards (NAAQS). Tamil Nadu, which is in southern India, Gujarat and Maharashtra in western India have concentrations of particulate matter that are above norms prescribed in NAAQS-2009. High particulate emissions makes meeting NAAQS very challenging. The pilot systems for the three included states will cover 1,000 facilities. SPCBs determine the precise eligibility criteria for facilities. The Maharashtra pilot programme encompasses the cities of Aurangabad, Tarapur, Chandrapur, Jhalna, and Kohlapur. Selected unitsmust be of medium or large size, be high emitters of PM (particulate matter), and have at least one stack suitable for Cost Effective Monitoring System (CEMS). The Tamil Nadu pilot system will encompass the cities of Ambattur, Chennai, Maraimalai, Sriperumpudur, and Tiruvallur. Covered facilities must lie within a 50 km radius of Chennai City, have at least one CEMS-suitable stack and be of medium or large size. The Gujarat pilot ETS will encompass the cities of Surat, Vapi, and Ahmedabad. A covered industry must lie within a 20 km radius of one of these three cities, be a high emitter of PM, and have at least one CEMS suitable stack. The programme is modelled after successful past and present cap-and-trade mechanisms, namely the US SO2 trading programme, the Chilean offsets trading programme for suspended particulates, and EU ETS for GHGs. A country-wide report from February 2011 estimates the total budget for design and roll-out of this ETS to be INR 360 crore (USD 57.7 million); the single largest cost, comprising over 95 per cent of the overall budget, is installation and maintenance of CEMSs.

Renewable Energy Credit (REC) Trading System

India's REC trading system was launched in November 2010, and the system's primary purpose is to promote renewable energy even in regions that have low potential for renewable power generation.

To comply with their RPOs or profit from a surplus of RECs, covered entities may trade RECs either within or across states. The two key actors in this scheme are eligible entities and obligated entities. The eligible entities are the generators of renewable power and the obligated entities are the purchasers of this power or the utilities. Purchasing RECs is treated as the consumption of the corresponding quantity of renewable power and can thus be used towards fulfilling the Obligated Entities' RPO. There are two types of RECs, solar and non-solar. An amendment in the Electricity Act of 2003 required utilities to source at least 0.25 per cent of their RPO from solar power exclusively, starting 2012-13. Each REC represents one MWh of a covered type of renewable energy, solar, wind, small-scale hydro (capacity below 25 MW), biomass-based power, bio fuels, and municipal waste based power. As a result, facilities are able to meet their renewable energy targets even if the local climate is not well-suited for renewable energy generation. The REC system enables obligated entities to weigh the costs and benefits of achieving their renewable energy commitments by selling electricity from renewable sources or by purchasing RECs.

Regulation and Oversight

The Environmental Protection Act of 1986 and accompanying rules empower the MoEF to limit adverse effects of industrial activity. In addition, the State Pollution Control Boards (SPCBs) have the power to implement emissions trading systems within their jurisdictions. While an enabling framework exists for emissions trading, details such as monitoring standards must be elaborated before implementation. For the pilot ETS programmes in Tamil Nadu, Gujarat, and Maharashtra, MoEF introduces the ETS regulatory framework. The CPCB sets monitoring and reviewing standards. Then SPCBs implement monitoring and enforcement procedures. The legal framework for PAT is embedded in various sections of the Energy Act of 2003 (EA 2003). Legal mandates include:

i) Covered entities must submit a report of energy consumption to the Designated Authority of the State and to the BEE.

ii) Covered entities must establish an Energy Manager responsible for submitting annual energy consumption returns to the Designated Agencies and BEE.

iii) Covered entities must comply with prescribed energy conservation standards.

iv) Covered entities must purchase ESCerts to avoid defaulting on compliance obligations.

v) Designated Energy Auditors (DENAs) must monitor and verify covered entities' compliance.

vi) Covered entities must be able to receive ESCerts that can be sold on a market if the compliance obligation is exceeded.

vii) Non-compliant entities must pay INR 10 lakh (USD 15,710); and

viii) Covered entities must submit to regulation by BEE and process management by Energy Efficiency Service Ltd. (EESL).

The REC scheme is monitored and verified through the combined efforts of the CERC and the SERCs. To receive accreditation for the renewable energy projects, physical inspections and verifications are conducted by the SERC and then the CERC, after which projects receive a Certification of Accreditation and Certificate of Registration, respectively.

The CERC is also responsible for validating the transactions occurring on the trading platforms. The SERC ensures Obligated Entities comply with their RPO requirements. For obligations that have not been met, the SERC has the power to penalise the entities and enforce them to meet their RPOs through purchase of RECs.

Complementary Policies

A key step in terms of policy and climate action was the launch of India's first ever National Action Plan on Climate Change on 30th June 2008. This plan outlined existing and future policies addressing climate mitigation and adaptation. It also set up seven national missions that will run through 2017. India's goal for solar, which was approved in 2009, is to increase PV power production and solar thermal generation to 1,000 MW per year, as well as to strengthen research, development, and deployment of solar energy. The voluntary target of the solar mission was to increase solar capacity to 20,000 MW by 2020.52 This specific goal, as previously mentioned, has been superseded by a new target of increasing solar capacity to 100 GW by 2022.

Box 1.1

Emission Trading Schemes of India

Features

1. PAT is the first of its kind—a market system geared towards enhancing energy efficiency—in the developing world.
2. Few other countries, especially developing countries, have three national market-based environmental programs—PAT, REC, and CDM—that are actively reducing GHGs.

Challenges

1. India will need to build its capacity, namely improve its data collection and its supply of trained manpower, to implement an ETS effectively.
2. The transition from a PAT system to a well-functioning ETS will be challenging and require the Indian government to change its climate priorities away from increasing energy production, and move towards limiting greenhouse gases.
3. Non-compliance penalties are relatively weak so they could fail to incentivise compliance.

Applicability of Emission Trading Scheme

The application of cap-and-trade pollution markets has been very successful in practice, achieving the desired reductions in emissions reliably and at lower than-expected cost. This section briefly considers three different emissions trading schemes for three different air pollutants that have been implemented in different countries.

Sulfur Dioxide, United States

An extremely successful trading scheme that reduced sulfur dioxide emissions and the related acid rain problem at a surprisingly low cost to firms.

Purpose: The 1990 Clean Air Act Amendments established the first large-scale system to reduce pollution through tradable emissions permits. (Schmalensee et al. 1998) The purpose of the program was to reduce emissions of SO2 from power plants to stop acid rain and for this reason it is also sometimes known as the Acid Rain programme. The overall goal of the program was a reduction in annual SO2 emissions by 10 million tonnes below 1980 levels.

Emissions Cap: This program was set at about half of the level of emissions from the same set of power plants in 1980, beginning in 1995. Phase I subjected only the 263 dirtiest power plants in 21 states to the cap and Phase II, beginning in the year 2000, applied to over 2,000 fossil-fuel burning electric generating units, virtually all such units in the United States. Each participating unit is required to hold an amount of permits, or allowances, at least equal to its annual emissions. If a plant emits 5,000 tonnes of SO2, it must hold 5,000 permits to do so. Regardless of the number of permits a plant holds, it may not exceed upper limits set by the same act to protect public health. (Environment Protection Agency 2009) The overall level of emissions was set on a per-unit basis based on baseline heat input of the electricity generating units. Permits were allocated for each year beginning in 1995. The EPA allocated allowances at an emission rate of 2.5 pounds of SO2/mm Btu (million British thermal units) of heat input, multiplied by the unit's baseline mm Btu (the average fossil fuel consumed from 1985 through 1987).

Implementation: Existing generating units were given, free, a number of permits based on the formula above at the beginning of the program. Note that the determination of the cap and the allocation of the units are in principle separate; the EPA could have set the overall cap using the formula above but then allocated permits using an auction.

Trading: After the initial allocation, new units or units that need additional permits could obtain them from two sources: purchasing from other or at auctions run by the EPA. The purpose of the auction was to provide clear information about prices, but in practice most trading occurred privately.

Trading enables units to achieve emissions reductions at low cost without changing the overall level of pollution set by the cap. Many units have actual emissions rates higher or lower than they would have been under a fixed emissions standard, indicating that units used the flexibility of emissions trading to achieve compliance at a low cost. For some units, the best way to comply is to reduce emissions directly, for others it is to pay for the right to emit more by buying credits from units that were able to abate emissions at a lower cost. The price of emissions urged each unit to seek out low-cost means of abatement while the cap ensured that overall emissions fell.

Monitoring: Emissions are monitored continuously at affected units, at an average annual cost of $124,000. Continuous emissions monitoring is an essential foundation for trading, which requires that every ton of unit emissions be recorded. Units required to monitor SO2 emissions in pounds per hour must use both an SO2 pollutant concentration monitor and a volumetric flow monitor. Both concentration and flow are required because the permit applies to the total mass of the pollutant emitted. In addition to this basic equipment the unit must have a Data Acquisition and Handling System (DAHS) to process and record readings. The monitoring protocol includes conservative formulas for filling-in missing data that penalise units for monitoring downtime. The EPA also has monitoring protocols for NOX and CO_2 emissions. The number of permits held by each unit is tracked in an Allowance Management

System (AMS). Parties may trade permits privately but must notify the EPA when permits acquired through these trades are going to be used to cover emissions. Affected units must deliver to the EPA valid permits sufficient to cover a year's emissions by 1 March of the year following.

If a unit does not have enough permits it must surrender any permits for the following year as excess emission offsets and pay a penalty for the present year's excess. (Federal Register 1997) The penalty is the number of excess tons emitted times a base fine of \$2,000 per tonne, an excess charge that is adjusted annually for inflation. The excess charge for 1997 was \$2,525 per ton. As permit prices in this year were little above \$100 per ton the excess charge is enormous and in practice all units complied by holding sufficient permits.

Outcomes: The effect of the programme was a large reduction in emissions at a lower cost than anticipated. In 1995, the first year of the program, total emissions fell from 8 million tons to less than 5 million tons. Generating units achieved these reductions by using lower-sulfur coal and by using scrubbers to remove SO2 from stack gases. The estimated savings to firms from using a trading scheme instead of fixed regulations was around \$225 million to \$374 million per year (Ellerman et al. 1997).

Total Suspended Particulates, Chile

An inconsistently implemented trading scheme that nonetheless may have helped push firms to use cleaner fuels.

Purpose. The Emissions Offsets Trading Program was introduced in 1992 to reduce airborne particulate emissions from stationary industrial sources in Santiago, Chile (Montero and Sanchez 2002). Standards for particulate matter less than 10 nanometers in diameter (PM10) had been consistently exceeded in Santiago since the 1970s and these high levels have been related to mortality and respiratory disease (Ostro et al. 1999).

Emissions Cap. At the time the regulation was issued, existing sources received daily emissions rights in proportion to a pre-determined baseline emissions rate that was uniform across existing sources. The Program aimed to reduce emissions by 80 per cent, a figure chosen to achieve daily ambient air-quality standards 95 per cent of the time. The cap applied to sources with a flow rate greater than 1,000 m3 per hour, of which there were a total of 680 in 1993. Both existing and new sources are subject to a maximum cap that cannot be exceeded regardless of the number of permits held by that source. This overall cap was usually about twice the level of emissions permits granted a source. In addition, large stationary sources were liable to be shut down during episodes of severe ambient air quality, again regardless of the number of permits held.

Implementation: Existing sources were granted emissions permits free and in perpetuity, or "grandfathered" in. Any new sources arising after the program began were obligated to purchase rights from existing sources. This difference made the introduction of the program very successful in getting existing but unknown point sources to identify themselves in order to secure emissions rights. Because unknown sources emerged to secure rights, however, and rights were based on a per unit emissions capacity formula, the overall cap ended up less stringent than expected.

Trading: Rights to emit are based on annual reviews of capacity. Therefore the only requirement is that at the time of this annual monitoring the unit holds enough capacity rights to cover its current capacity. In practice, trading was relatively thin—not many emissions capacity rights changed hands. This low trading is an indication that the market did not realise its full potential for cost savings, as capacity rights have stayed with their initial holders, which may not be the firms to which these rights are most valuable. Low trading appears to have been a reaction to uncertainty about future regulations and transactions costs in the permit market.

Monitoring: The authorities monitored not aggregate emissions but emissions capacity, which was estimated based on field visits to measure source size and fuel type. This is a rough estimate of the potential to produce emissions, which may serve better for particulate matter than for other pollutants, but does not allow for the ability to reduce emissions via better combustion control or other channels that do not change capacity but may reduce emissions. Another limitation of this monitoring method is that it excludes process sources that produce emissions but are not comparable in terms of capacity to boilers. That is, the emissions capacity of each unit is based on estimated flow volume and a uniform emissions concentration quota, but no such quota could be applied to diverse industrial process sources, limiting the scope of the system. For these reasons it is preferable, with modern technology, to monitor aggregate pollutant emissions directly.

Outcomes: Total TSP capacity (not emissions per se) was below the number of capacity rights by 1997. Industry switched to cleaner fuels, such as natural gas, during this period, which appears responsible for most of the decline. There are not good estimates of the cost of these changes but it seems that the establishment of the permit system accelerated this switch and created a response from gas supply.

Carbon Dioxide, European Union

A well-functioning scheme that established thriving markets for carbon and reduced emissions but may have set initial emissions too high.

Purpose: The European Union introduced a CO_2 emissions trading scheme, the EU ETS, starting in 2005 in order to limit emissions of carbon dioxide that contribute to global climate change. The scheme applies to all EU member states and, within these states, to companies that generate electricity or are in energy-intensive industrial sectors.

Emissions Cap: As the EU ETS is an international program the allocation process was conducted at the level of each country. Each member country was responsible for developing a National Action Plan (NAP), subject to review by the European Commission, that details the quantity of emissions each state expects and how it will allocate its permits to individual emitting units. Unit-level allocations were typically based on historical emissions and covered a total of about 12,000 units emitting 46 per cent of the EU total CO_2 emissions. (Robinson and Watanabe 2005) EU member states collected emissions data on years up to 2004 to establish NAPs that set expectations for what emissions would be over 2005-2007 and were used for allocating permits to the units included in the scheme. This data was largely supplied voluntarily by facilities that would be receiving allocations and therefore had an incentive to distort emissions upwards. (Buchner and Ellerman 2009) There was no single, Europe-wide allocation target and member states tended to allocate to their own units generously relative to both historical emissions and Kyoto protocol targets (Azar et al. 2005).

Implementation: The implementation of the scheme took place in two phases. First phase was from 2005-2007 and the second from 2009-2012. Member states were required to give away 95 per cent of the permit allocations during the first phase and 90 per cent during the second phase to units free of charge.

Trading: The market for permits to emit carbon dioxide has been open and active since its inception. Carbon prices have generally ranged from about €10-30 with considerable movement in that range. Not only units but also many third-parties participate in the carbon market, and permits from the Kyoto protocol, which represent reductions in carbon emissions from outside the EU, are to some extent trade able with the EU ETS as well.

Monitoring: Units registering for permits are required to have a system to monitor and record emissions in place. During the scheme, units are then required to surrender enough permits by 30 April of each year to cover emissions from the previous year. Units that do not have enough permits have to pay an excess emissions penalty of €40 per tonne of CO_2 during the first phase of the scheme and €100 per tonne in the second phase. These penalties constituted big premiums over prevailing prices during the first years of the scheme.

Outcomes: The EU ETS lowered carbon emissions about 3.4 per cent below the business-as-usual case. (Buchner and Ellerman 2008) Emissions trading within and across countries did a good job in matching units with excess allocations to those with deficits (Ellerman and Trotignon 2008). Initial allocation targets were generous, but the positive emissions price indicates the system did have bite and will work to achieve further reductions in the future.

Conclusion

All the countries across the globe are working on reducing Green House Gas Emission. As a part, various countries viz., Europe, New Zealand, Australia, Kazakhstan, Switzerland and India have been implementing ETSs. The countries are integrating cap-and-trade systems into their climate change policies. They have levied these schemes for complying national CO_2 levy on heating, industrial process, and transport fuels. Australia started collecting tax from 500 companies that are emitting carbon around 5 percent per year. New Zealand ETS is the first to place a cap on emissions from deforestation. China accounts for almost a quarter of global GHG emissions. India is currently the third largest polluting country in the world. India's National Action Plan on Climate Change which specifies eight national objectives for 2017 that centre on improving energy efficiency, solar technology, sustainable habitats, water, Himalayan ecosystems, "green India", agriculture and strategic knowledge. India's PAT initiative is in first phase i.e., from 2012-2015, which is considered as a test phase. PAT covers facilities in nine sectors viz., thermal power, iron and steel, cement, fertilisers, textiles, aluminium, pulp and paper, Chlor-Alkali and railways. The application of cap-and-trade pollution markets has been very successful in practice, achieving the desired reductions in emissions reliably at lower than expected cost.

References

Argus Media Ltd (2012). *India Initiates Pilot Emissions Trading Schemes*. 16 April. Available at: *http://www.argusmedia.com/pages/NewsBody.aspx?id=793879&menu=yes*

Azar, C., M. Grubb and U.M. Persson (2005). "Allowance Allocation in the European Emissions Trading System: A Commentary", *Climate Policy* 5(1): 127-36.

Buchner, B. and A.D. Ellerman (2006). "Over-allocation or Abatement? A Preliminary Analysis of the EU ETS Based on 2005 Emissions Data", *Environmental and Resource Economics* 41(2): 267-87.

Climate and Development Knowledge Network (2013). *Creating Market Support for Energy Efficiency: India's Perform Achieve and Trade Scheme*. January.

Ellerman, A.D., R. Schmalensee, P. Joskow, J.P. Montero and E. Bailey (1997). *Emissions Trading Under the U.S. Acid Rain Program: Evaluation of Compliance Costs and Allowance Market Performance*. MIT Center for Energy and Environmental Policy Research.

Ellerman, A.D. and R. Trotignon (2008). "Compliance Behavior in the EU ETS: Cross Border Trading, Banking and Borrowing", *MIT Center for Energy and Environmental Policy Research* No. 08-12.

Environmental Protection Agency (2009). *Continuous Emissions Monitoring Fact Sheet*. Available at *http://www.epa.gov/airmarkt/emissions/continuous-factsheet.html* (last checked 28 July 2010).

Federal Register (1997). *Federal Register* 62(194), 7 October.

Fankhauser, S. (2011). "Carbon Trading: A Good Idea Is Going Through a Bad Patch", *The European Financial Review* 32-35.

Government of India (n.d.). *Market Friendly Emissions Scheme*. Ministry of Environment and Forests.

Jindal, R.N. (2011). *A Case for a Pilot Emissions Trading Scheme in India. MoEF, India*. 19 September. Available at: *http://www.hks.harvard.edu/var/ezp_site/storage/fckeditor/file/pdfs/centersprograms/ centers/mrcbg/programs/ssp/docs/events/conferences/2011/harvard/Jindal_Harvard_Sust_Science_Forum_110919.pdf*

Liu, C. (2011). "China Moves Toward Carbon Emissions Trad¬ing to Improve Energy Efficiency and Competitiveness", ClimateWire in *The New York Times*. Available at *http://www.nytimes.com/cwire/2011/01/10/10climatewire-china-moves-toward-carbon-emissions-trading-81996.html*

Montero, J. P., J.M. Sanchez and R. Katz (2002). "A Market-Based Environmental Policy Experiment in Chile", *The Journal of Law and Economics 45*.

Olivier, J.G.J., G. Janssens-Maenhout, J. Peters and J. Wil¬son (2011). *Long-term Trend in Global CO_2 Emissions: 2011 Report.*

Ostro, B., G. Eskeland, J.M. Sanchez and T. Feyzioglu (1999). "Air Pollution and Health Effects: A Study of Medical Visits Among Children in Santiago, Chile", *Environmental Health Perspectives 107(1).*

Pechan, E.H. (1995). *The Acid Rain Database Version 1 (ARDBV1). E. H. Pechan Associates, Contract no. 68-D3-0005*. Prepared for the U.S. Environmental Protection Agency's Offi ce of Atmospheric Programs, Acid Rain Division, Washington, D.C.

Robinson, G. and R. Watanabe (2005). "The European Union Emissions Trading Scheme (EU ETS)", *Climate Policy* 5(1): 10-14.

Schmalensee, R., P. Joskow, A.D. Ellerman, J.P. Montero and E. Bailey (1998). "An Interim Evaluation of Sulfur Dioxide Emissions Trading", *Journal of Economic Perspectives* 12(3): 53-68.

Stavins, R. (2007). Proposal for U.S. Cap-and-Trade System to Address Global Climate Change: *A Sensible and Practical Approach to Reduce Greenhouse Gas Emissions*. The Hamilton Project.

———. (2003). Experience with Market-based Environmental Policy Instruments. *Handbook of Environmental Economics I*. pp.355-435.

Steckel, J.C., M. Jakob, R. Marschinski and G. Luderer (2011). "From Carbonization to Decarbonization? Past Trends and Future Scenarios for China's CO2 Emissions", *Energy Policy* 30(6): 3443-55, June.

The Economic Times (2011). " India to Unveil Emissions Trading Scheme on February 1", *The Economic Times,* 27 January. Available at: *http://articles.economictimes.indiatimes.com/2011-01-27/news/28430106_1_emissions-trading-pollution-control-emission-levels*

Upadhyaya, P. (2010). "Is Emission Trading a Possible Option for India?", *Climate Policy* 5(10): 560-74, June. Available at: *http://www.tandfonline.com/doi/pdf/10.3763/cpol.2010.0105*

2

SRILATHA VANI and A. LALITHA

Constraints Faced in Practising Organic Farming among Farmers of Khammam District

Introduction

The conventional farming had helped India not only to produce enough food for own consumption but also generated surpluses for exports. However, the increasing population and income will lead to further increase in demand for food and also for raw materials for industry. The modern system of farming, it is increasingly felt, is becoming unsustainable as evidenced by declining crop productivities, damage to environment, chemical contaminations, etc. The necessity of having an alternative agriculture method which can function in a friendly ecosystem while sustaining and increasing the crop productivity is realised now. Organic farming is recognised as the best known alternative to the conventional agriculture.

The progress of organic agriculture in India is very slow. We are able to convert only 41,000 ha of area so far, which is a mere 0.03 per cent of the cultivated area. In India, only less than 2 per cent farmers are practising organic farming. Organic farming growth is losing ground and existing farmers are withdrawing from the practice.

The growth of organic farming is found to be losing ground in spite of the efforts being made by the well wishers of people and environment, government and agriculture universities. A few conventional farmers are entering organic farming but equally existing organic farmers lost their hope in organic farming and started practising both organic farming and conventional farming practices. Hence, a study has been taken with selected sample organic farmers to know about the constraints faced by the farmers.

In India, the organic practices and organic farming are practised by a very low agrarian population. Hence the present study has been undertaken to evaluate various factors affecting the growth of organic farming. The work carried out based on the major objectives viz., constraints in adoption of organic farming practices and find out solutions to overcome the problems originating in organic agriculture.

Methodology

This study was conducted in Khammam district of Andhra Pradesh (AP). Out of 42 *mandals* of the district, 3 *mandals* namely Pinapaka, Wyra and Mudigonda were selected for the purpose. From each selected *mandal,* 40 farmers were selected randomly as respondents who are aware of organic farming and practising organic farming. In the present study, constraints are defined as problems/difficulties for organic farming that exist and as foreseen in adopting organic crop production practices. Data collection was done by using schedules which consists of open-ended questions. For the collected constraints, frequency and percentages were calculated and then ranking was given. The constraints were classified into constraints encountered in organic farming and constraints of using organic fertilisers.

Results and Discussion

From the results in Table 2.1, it can be observed that problems in organic products certification was ranked highest by the organic farmers. The other constraints identified by them in the descending order were: reduced yields in the initial years of organic farming, increased weed cover, marketing problem, lack of consumers understanding about organic products, labour shortage, no financial help from government, lack of domestic market information, no training programme to promote organic farming and lack of knowledge in some of the concepts of organic farming. The results are in line with that of Siddaraju and Rajendran (2006) and Sivaramana (2007).

Table 2.1

Constraints Encountered in Organic Farming

S.No.	Problems Encountered	(%)	Rank
1.	Lack of domestic market information	44.40	X
2.	Problem in organic products certification	100.00	I
3.	Lack of knowledge in organic farming	41.66	IX
4.	Reduced yields in the initial years of organic farming	83.33	II
5.	Lack of training programmes for promoting organic farming	54.16	VIII
6.	No financial help from government	60.00	VII
7.	Labour shortage and high labour cost	66.67	VI
8.	Increased weed cover	76.66	III
9.	Marketing problem of organic products	75.00	IV
10.	Lack of consumer understanding about organic farming	70.83	V

From Table 2.2, it is evident that majority (75%) of the organic farmers faced problem of low availability of certified organic inputs followed by unavailability of organic inputs, low effectiveness of organically allowable inputs and methods, lack of skills regarding preparation of organic inputs, high cost of handling bulky organic manures, etc.

Table 2.2

Problems of Using Organic Fertilisers

S.No.	Problems Encountered	(%)	Rank
1.	Low availability of certified organic inputs	75.00	I
2.	Slow decomposition of compost	35.00	VII
3.	High cost of handling bulky organic manures	51.66	V
4.	Unavailability of organic inputs	73.33	II
5.	Limitation in availability of green leafy manures	41.67	VI
6.	Lack of skills regarding the preparation and application of organic inputs	56.67	IV
7.	Low effectiveness of organically allowable inputs and methods	66.67	III

Conclusion and Recommendations

Generally, organic farming could be successful through increased visits by technical persons, holding workshops or short courses for farmers and supplying farmers with inputs like seed for organic agriculture. Farmers should get adequate training on preparation and post-emergence application of organic fertiliser. Markets for organic crops should be opened to encourage farmers to appreciate the benefits of organic farming. The returns from organic farming should also be quantified for easy comparison with returns from conventional crops. To counteract the problem of shortage of organic material, the application of organic fertilisers by foliar methods could reduce the quantities required per unit area, especially during the vegetative phase. More research is also required on the application rates and amounts to be soaked in water for foliar application. In addition, research is needed to compare efficiencies of organic fertilisers that are developed from different kinds of living organisms. Success in the implementation of organic farming techniques in the smallholder sector would be a great step towards making agricultural production sustainable.

References

Siddaraju, V.G. and S. Rajendran (2006). "Organic Farming System: Issues and Concerns", *Kurukshetra* 54(9): 20-23.

Sivaramana, K. (2007). "Organic Farming: Concepts, Issues and Strategies with Reference to Spices", *Green Farming* 1(3): 20-24.

3

DAKURI ARJUN, V. BHASKER and
DAKURI CHIRANJEEVI

Treatment Methods for Textile Wastewater

A Solution to the Environmental Pollution

Introduction

Textile is one of the important and leading sector which produces excessive waste. The textile dyeing industry consumes large quantities of water and produces large volumes of wastewater from processing *viz.*, desizing, scouring and mercerisation, dyeing and finishing processes. The dyed wastewater contains residual dye stuff, chemicals and other contaminants, which requires proper treatment before being released into the environment. The following are the main reasons of pollutants in the wastewater:

1. The natural impurity in fibres.
2. The chemical materials that are used in processes.

The characteristics of textile wastewater are high chemical oxygen need, high biological oxygen need, high temperature, high PH, solid materials, phenol, sulphur and the colours caused by different dyes. The contemporary challenge for the textile industry is to modify production methods, so they are more ecologically friendly at a competitive price, by using safer dyes and chemicals and by reducing cost of effluent treatment. There are three ways to reduce pollution:

1. New technologies.
2. Effective treatment of effluent.
3. Recycling waste several times over before discharge.

The various methods used for the textile effluent are discussed in this paper including conventional and novel techniques (Babu et al. 2007; Sule and Bardhan 1999).

Textile Operations

There are many steps involved in the conversion of raw fibres into finished product. The main stages in the conversion include yarn formation, fabric formation, wet processing and textile fabrication. Following are the steps involved in the wet processing operation.

Desizing

Sizing operation is used to impart the weave ability in the warping section; the presence of sizing ingredients in the fabric hampers processes i.e., dyeing, printing and finishing. Therefore it is very essential to remove the added size by a process called desizing. In this process, starch is converted into water-soluble products either by hydrolysis or by oxidation. In general, about 50 per cent of the water pollution is due to wastewater from desizing, which has a high BOD (biological oxygen demand) that renders it unusable. This high BOD can be reduced by using enzymes which degrade starch into ethanol rather to anhydroglucose. The ethanol can be recovered by distillation for use as a solvent. The other sizing ingredients such as polyvinyl acrylic and acrylic resins can be recovered by nano-filtration and ultra-filtration techniques (Batra 1985).

Mercerisation

This process is carried out to impart lustre, strength and improve dye uptake of the material. In this process, cotton material is treated with a strong solution of sodium hydroxide (NaOH) (about 18-24%) and washing off the caustic after 1-3 mins under tension to avoid the shrinkage. The large concentrations of NaOH in the wash water can be recovered by membrane techniques. Zinc chloride (ZnCl) as an alternative method leads to an increase in the weight of fabric and dye uptake. Moreover, the process is ecologically friendly and does not require neutralisation by acetic or formic acid (Kareem et al. 2006).

Bleaching

Presence of natural colour matter in the yarn imparts a creamy appearance to the fabric, it is necessary to decolourise the yarn by bleaching. Hypochlorite is one of the oldest industrial bleaching agents. Past few years, hypochlorite is being replaced by peracetic acid, which decomposes into oxygen and acetic acid, it is completely biodegradable. One of the advantages of peracetic acid is higher brightness values with less fibre damage. Slokar and Majcen (1997) reported the one-step preparatory process for desizing, scouring and bleaching has helped to reduce the volume of water (Rott and Minke 1999).

Dyeing

Application of chemical pigments to the fibre, yarn or fabric to impart colour is called dyeing. The presence of chromophore and auxochrome groups in the dyes is the main source for the colouration of textile material, which also cause pollution. Large amount of water is required for dyeing process based on the amount fabric to be dyed. Once the dyeing operation is over, the various treatment baths are drained, including the highly coloured dye bath, which has high concentrations of salt and organic substances (Chakraborty et al. 2005).

Printing

It is generally defined as localised dyeing, i.e., dyeing. In this process, colour is applied in the form of a thick paste of the dye. The fixation of the colour in printing is brought about by a suitable after-treatment of the printed material. The process of printing produces hydrocarbon effluents that must be removed before they reach the atmosphere. In India, a majority of textile printing units prefer to use kerosene in printing because of the brilliant prints and ease of application. Air-laden kerosene is harmful to human beings, as well as to the flora and fauna. Therefore, it is imperative that as much kerosene as possible to be recovered from the exhaust pipes of the printing industry (Babu et al. 2007).

Finishing

This method is used to improve specific and functional properties viz., water repellent, flame retardant, easy care, wrinkle resistant and bacterial resistant in the finished fabric and involves the use of a large number of finishing agents for softening and cross-linking, which contribute to water pollution (Babu et al. 2007). Figure 3.1 shows various steps involved in textile processing industry.

Figure 3.1

Process Flow Diagram for Textile Wet Processing

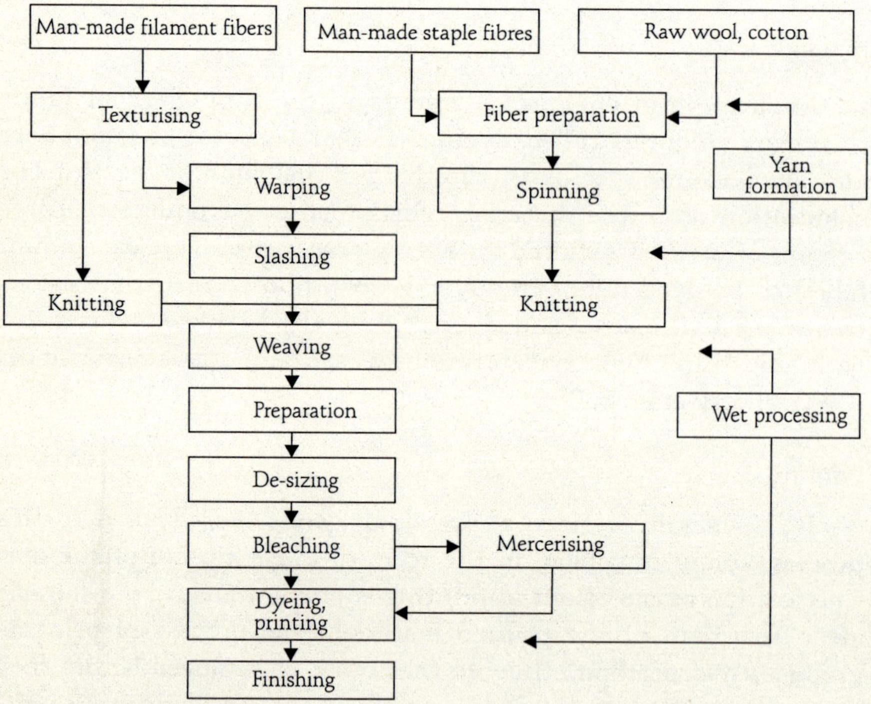

Source: Babu et al. 2007.

Environmental Impact

Basic dyes and diazo direct dyes have the highest toxicity among 4,000 dyes as per the Ecological and Toxicological Association of the Dyestuffs Manufacturing Industry (ETAD) in the year 1974. The presence of azo- and nitro-compounds in the structure of organic dyes has the possibility to generate aromatic amines if the textile effluent is incompletely treated. Aromatic amines are reported to be capable of inducing cancer and tumours in human. The toxicity of textile effluents was measured using Daphnia magna neonates for 24-48 hours of exposure. They found that all the wastewater samples were toxic in terms of lethal concentration and exhibited very high toxicity with acute toxicity unit level between 2.2 and 960. The release of these metals and the persistent dyes structure with toxicity properties may cause negative impacts to human health and the ecosystem. Many adverse effects from the consumption of dyes with respect to their carcinogenic genotoxic, mutagenic and teratogenic properties in animal studies have been reported. Apart from this, the non-biodegradable nature of organic dyes and their high colour intensity are able to reduce aquatic diversity by blocking the passage of sunlight through the water and some of them are capable of causing irritations to the skin, eyes, allergic dermatitis and respiratory tract if swallowed (Pang and Abdullah 2013).

Effluent Treatments

The wastewater from the dye house is generally multicoloured leftover dye stuff and other chemical ingredients. The depth of penetration of sunlight into the water environment decreases with increased effluent disposal, which in turn decreases photosynthetic activity and dissolved oxygen. The adverse effects can spell disaster for aquatic life and the soil. The following are the effluent treatments including chemical, biological and physical treatments (Babu et al. 2007).

Adsorption

It is an equilibrium separation process, which has high potential in wastewater treatment application. Activated carbons are used in this technology, which have capable of adsorbing different dyes with high adsorption capacity. However, this method is very expensive and the need for activated carbon regeneration causes limit is success in wastewater treatment and this process will yield waste sludge which causes disposal problem (Copper 1995).

Ion Exchange

This technique can be used for the removal of undesirable anions viz., acid, direct and reactive dyes, and cations (basic dyes) from wastewater. However, the removal of non-ionic dyes (disperse dyes) is very poor. The regeneration of ion exchange resin bath is required once the exchange capacity of ion resin is saturated, the use of expensive organic solvent to regenerate spent ion exchange resin is unavoidable. The advantage of this technique includes the recovery of adsorbent upon successful regeneration, reclamation of solvent after use, and the effective removal of soluble dyes (Babu et al. 2007).

Filtration

Filtration technologies i.e., microfiltration, ultrafiltration, nanofiltration and reverse osmosis are used for treating textile effluents. Among these, microfiltration and nanofilatration are the most economically competitive alternatives because of larger pore size for its building modules. However, some of the disadvantages include practical limitation to only smaller wastewater flow rate, high cost of membrane system, high working pressure to push the wastewater flow through membrane and inability to reduce dissolved solid content. Besides, the life time of membrane is relatively short due to frequent clogging on the membrane pores. Thus, this method requires frequent cleaning and replacement of the modules to maintain its effectiveness in removing organic dyes (Babu et al. 2007).

Reverse Osmosis

This technique retains maximum amount of ionic compounds and produces a high quality of permeate. Decolouration and Elimination of chemical auxiliaries in dye house wastewater can be carried out in a single step by reverse osmosis. Reverse osmosis permits the removal of all mineral salts, hydrolysed reactive dyes and chemical auxiliaries (Babu et al. 2007).

Nanofiltration

A combination of adsorption and nanofiltration can be adopted for the treatment of textile dye effluents. The adsorption step precedes nanofiltration, which decreases concentration, polarisation during the filtration process, which increases the process output. A major problem is the accumulation of dissolved solids, which makes discharging the treated effluents into water streams impossible. Nanofiltration treatment as an alternative has been found to be fairly satisfactory. The technique is also favourable in terms of environmental regulation (Chakraborthy et al. 2003).

Ultrafiltration

Ultrafiltration enables elimination of macromolecules and particles, but the elimination of polluting substances, such as dyes, is never complete. Even in the best of cases, the quality of the treated wastewater does not permit its reuse for sensitive processes (Babu et al. 2007).

Microfiltration

Microfiltration is suitable for treating dye baths containing pigment dyes, as well as for subsequent rinsing baths. The chemicals used in dye bath, which are not filtered by microfiltration, will remain in the bath. Microfiltration can act as a pre-treatment alternative before further treatment using nanofiltration or reverse osmosis to satisfy the reusability criteria for textile industry (Malack and Anderson 1997).

Biological Treatment

This technique may involve the use of microorganisms in both conditions i.e., either aerobic or anaerobic to degrade organic dyes by using fixed or suspended growth systems. The possible microorganisms used for biodegradation dyes are fungi, bacteria, yeasts, algae and biosorption by living or dead microbial biomass. The microorganisms used in this process are capable to metabolise the biodegradable organic to be converted into water, carbon dioxide and energy for their growth and reproduction. However, this is not effective for treating of raw effluent from textile industry. Pretreatment is often required due to the low biodegradability nature of textile dyes wastewater. In aerobic conditions, enzymes secreted by microorganisms in wastewater break down the organic compounds. The ability of white-rot fungi to mineralise organic dyes is attributed to their highly oxidative and non-specific ligninolytic enzymes including lignin peroxidase, manganese peroxidase and laccase. Bacteria decolourisation is usually faster than fungal decolourisation (Babu et al. 2007, Pala and Tokat 2002).

Chemical Treatment

This technique uses coagulation/flocculation and chemical oxidation processes. The former method is a process commonly used for removing colloidal and other suspended particles from wastewater. Chemical oxidation has attracted more attention as it does not generate any secondary pollutant. Chemical oxidation processes generally involve the use of chemical reagents or the consumption of energy (UV or ultrasonic) without the need of microorganism for the reactions to proceed (Babu et al. 2007).

Coagulation and Flocculation

Coagulation and flocculation processes are among the most efficient methods which are widely employed in textile industry wastewater treatment as they are efficient for a short detention time and simple to operate. The principle of coagulation/flocculation process is to destabilise colloidal particles by charge neutralisation and promoting collision between neutralised particles, resulting in cohesion, floc growth and eventual sedimentation for effective filtration. However, coagulation and flocculation processes often lead to toxic sludge handling and disposal problems. The most popular coagulants in wastewater treatment application are aluminium sulfate or alum, lime and ferric salt such as ferric sulfate or ferric chloride. Although this process effectively eliminates insoluble dyes, its value is doubtful because of the cost of treating the sludge and the increasing number of restrictions concerning the disposal of sludge (Socha 1991).

Chemical Oxidation

Chemical oxidation is used to degrade the organic dyes due to its simplicity and effectiveness for destruction of organic compounds with non-biodegradable property (refractory), toxic or inhibitory to microbial growth and even inorganic compounds (Babu et al. 2007).

Adsorption on Powdered Activated Carbon

The adsorption on activated carbon without pretreatment is impossible because the suspended solids rapidly clog the filter. Therefore, it is feasible to use a combination of flocculation–decantation treatment or a biological treatment which permits a reduction of suspended solids and organic substances, as well as a slight reduction in the colour, but the cost of activated carbon is high (Babu et al. 2007; Rozzi et al. 1999).

Electrochemical Process

Electrochemical techniques are used widely for the treatment of wastewater from dyehouse. It has been applied to effectively remove acids, as well as dispersed and metal complex dyes. The removal of dyes from aqueous solutions results from adsorption and degradation of the dye-stuff following interaction with iron electrodes. If metal complex dyes are present, dye solubility and charge are important factors that determine the successful removal of heavy metals. pH maintenance is very important in this process to maximise dye insolubility (Babu et al. 2007).

Ozone Treatment

Ozone technique is now used in the treatment of industrial effluents and water treatment. Ozone especially attacks the double bonds that bestow colour which leads to the de-colourisation of wastewater, and ozone alone does not lead to a significant reduction in chemical, oxygen demand (COD). Apart from this installation of ozonation plants can entail additional costs (Babu et al. 2007, Stephan and Marco 2004).

Conclusion

This paper reviews various methods used for the textile effluents and to control the environmental pollution. Convention methods used for the textile wastewater treatment i.e., physical, biological, chemical and membrane are discussed. The membrane systems are among the best alternative methods that can be adopted for large-scale eco-treatment. A combination method involving adsorption followed by nanofiltration has also been summarised.

References

Babu, B. Ramesh, A.K. Parande, S. Raghu and T. Prem (2007). "Cotton Textile Processing: Waste Geneartion and Effluent Treatment", *Journal of Cotton Science* 11: 141-53.

B. Tan, T. Teng and A. Omar (2000). "Removal of Dyes and Industrial Dye Waste by Magnesium Chloride", *Water Res.* 34: 597-601.

Batra, S.H. (1985). "Other Long Vegetable Fibres: Abaca, Banana, Flax, Ramie and Coir", *Handbook of Fibre Science and Technology* IV: 15-22.

Chakraborty, S.S., J.K. Basu and S. Das (2005). "Treatment of a Textile Effluent: Application of a Combination Method Involving Adsorption and Nanofilitration", *Desalination* 174: 73-85.

Copper, P. (ed.) (1995). *Colour in Dyehouse Effluent Textile Executive.* Society of Dyers & Colourists.

Chakraborthy, S., M.K. Purkait, S. Das and J.K. Basu (2003). "Nanofiltration of Textile Plant Effluent for Colour Removal and Reduction in COD, Purif", *Technology.* 31: 141-51.

Gaeher, F., Hermanutz and Oppermann (1991). "Ozonation: An Important Technique to Comply with New Laws for Textile Wastewater Treatmen", *Water Sci. Technol* 30: 255-63.

Kareem M., A.K. Das and S.H. Lee (2006). "Treatment of Coloured Effluent of the Textile Industry in Bangladesh Using Zinc Chloride Treated Indigenous Activated Carbon, Anal", *Chim* 576: 37-42.

Malack, M.H. and G.K. Anderson (1997). "Use of Cross-flow Microfiltration in Wastewater Treatment", *Water Res.* 31: 3034-72.

Pala, A. and E. Tokat (2002). "Colour Removal from Cotton Textile Industry Wastewater in an Activated Sludge System with Various Additives", *Water Res.* 36: 2920-25.

Rott, U. and R. Minke (1999). "Overview of Wastewater Treatment and Recycling in the Textile Processing industry", *Water Sci. Technology* 40: 37-144.

Rozzi, A., F. Malpei, L. Bonnomo and R. Bianchi (1999). "Textile Wastewater Reuse in Northern Italy", *Water Sci. Technol* 39: 121-28.

Slokar Y.M. and M.A. Majcen (1997). "Methods of Decolourization of Textile Wastewater", *Dyes and Pigments*: 335-56.

Socha, K. (1991). "Treatment of Textile Efflients", *Textile Month* 12: 52-6.

Stephan, R. and R. Marco (2004). "Wastewater Treatment with Ozone", *Water Conditioning and Purification* 25-28.

Sule, A.D. and M.K. Bardhan (1999). "Objective Evaluation of Feel and Handle, Appearance and Tailorability of Fabrics Part-II", *Colourage* 46: 23-28.

Pang, Y.L. and A.Z. Abdullah (2013). "Current Status of Textile Industry Wastewater: A Review", *Clean Soil Air Water* 41: 751-64.

Section 2

Natural Resource Management for Sustainable Development

4

P. PADMANABHA RAO

Impact of Land Use on Urban Environment for Sustainable Development

A Case Study of Hyderabad

Sustainable city planning should aim at achieving social and environmental equity while improving the lives of the people. For this, we need to have a sustainable city form as well as provision and proper management of the services. Sustainable urban development refers to attaining social equity and environmental protection in urbanisation while minimising the costs of urbanisation. Thus, in order for a city or urban area to be sustainable it needs to produce and manage basic services like water, waste, energy, and transportation in a way that it conforms to the principles of sustainable development. With urbanisation comes the need to invest in infrastructure and improve the quality of life in our cities. Rapid urbanisation has not only outpaced infrastructure development, but has also brought in the downside of proliferating slums, increasing homelessness, growing urban poverty and crime, of relentless march of pollution, and ecological damage.

Flux of population in urban territories due to migration from rural areas coupled with rapid growth in population has disturbed the ecological balance. This process hampers the socio-economic sustainable development of any region. Expansion of suburban territory with encroachment in prime land is a matter of concern for all and in particular for the authorities associated with the urban planning and development. Spatial distribution of land use/land cover information and its changes is desirable for any planning, management and monitoring programmes at local, regional and national levels. For ensuring sustainable development, it is necessary to monitor the ongoing changes in land use/land cover pattern over a period of time. In order to achieve sustainable urban planning and to check haphazard development, it is necessary that authorities associated with the urban development generate such planning models so that every bit of the available land is used in most rational and optimal way. This requires the present and past land use/land cover information of the area and pattern of changes with respect to urban settlements and other local resources. There is a need for urban planning on a more sustainable basis to avoid environmental and social crises.

This paper aims at discussing some of the important issues relating to sustainable urban form that would lead to sustainable urban development of Hyderabad and land use change in the light of rapid urbanisation. Using Hyderabad Urban Agglomeration (HUA) 1971-2011 data, we have assessed how urbanisation and land use are interrelated and how assessment of physical growth could indicate sustainability of societal growth in future. Analysis of steady rise in regional population with a constant natural resource base purports unsustainable growth.

Hyderabad is one of the leading metropolitan cities of India. The city has under gone considerable socio-cultural change in the past. Today also it is undergoing rapid and dynamic changes due to accelerated phase of Industrialisation and urbanisation. These changes had a deep impact on the pattern and developmental trends in the city.

Objectives

The main objectives of the study are:

1. To study and analyse the population growth trend within the city and migration of people to the city in terms of development and its impact on land use changes.
2. To analyse the determinants of residential and land development for appraisal of land development problems and spatial development.
3. To study the urban land use dynamics and empirical investigation of some residential location models.

Methodology

The present study relies on a lot of secondary data to establish the growth trends in the city as well as the demographic details and those related to the amenities of the population. While the growth trends, etc., are shown with the help of graphs and tables, the other variable parameters are shown with the help of maps generated in the ARC GIS 9.2 environment.

The present study makes wide use of maps provided by various government agencies. Data were collected from secondary sources from published and unpublished sources from government records and others semi-government data sources such as Hyderabad Metropolitan Development Authority (HMDA), town planning authorities, Greater Hyderabad Municipal Corporation (GHMC) reports, Ministry of Urban Development and Planning reports and public institutions of Rangareddy district.

The Study Area

Hyderabad is the capital of Andhra Pradesh (AP), India. The city is one of the largest metropolitan of India with an area of 650 square kilometres (250 sq miles) with the population of 6,809,970 and further 7,749,334 living in the metropolitan area, making Hyderabad the fourth most populous city and the sixth most populous urban agglomeration in India. The city of Hyderabad with large migrants, most of whom occupied the core and suburbs of the city, catered to the needs of the growing population to maintain and improve the sanitary conditions of the city with ever-increasing people. The city has under gone considerable political and socio-cultural change in the past. Today also it is undergoing rapid and dynamic changes due to accelerated phase of industrialisation and urbanisation. These changes had a deep impact on the pattern and developmental trends in the city. The primary attribute of urban growth is population growth.

The city is located at 17°N to 17.5°N and 78°E to 78.5° E longitude, covering an area of 240 sq km. The Municipal Corporation of Hyderabad (MCH) area has 16 *mandals*. The areas covered by municipalities and the *gram panchayats* have 10 full *mandals* and

8 part *mandals*. Thus, Hyderabad Urban Development Authority (HUDA) area includes 26 full *mandals* and 8 part *mandals*. The MCH consists of 172.60 sq km, the 10 surrounding municipalities 418.58 sq km area and the 125 *gram panchayats* 1273.60 sq km together constitutes a total extent of 1864.87 sq km. In legal terms, this area was notified as the HMDA. In recent times certain changes took place because of the creation of special area development authorities like the Cyberabad Development Authority (CDA). But for overall and long-term metropolitan planning purpose, the above break up continues to cover the dimensions of a single metropolitan area.

The GHMC is formed by merging 12 municipalities and 8 *gram panchayats*, with the MCH in July 2007. The twin cities of Hyderabad and Secunderabad which have faced the full brunt of urbanisation and are near saturated so far as serving the civic requirements of the residents are the subject of study.

HMDA comprised of components: (1) GHMC (172.60 sq km), (2) 10 surrounding municipalities (416.58 sq km) and (3) 125 *gram panchayat* villages (1273.69 sq km) giving total area of 1864.87 sq km. However, Secunderabad Cantonment Board area of 40.17 sq km was never a part of HMDA. Eighty (80) per cent area of Hyderabad District, 20 per cent of Ranga Reddy and 2 per cent of Medak District constituted HMDA. In continuation of the series of development plans, CDA was set up in 2001 by carving out 52 sq kms from HUDA jurisdiction. The government also undertook Buddha Purnima Project Authority (BPPA) and Hyderabad Airport Development Authority (HADA) on the same lines.

Hyderabad city is one of the largest metropolitan of India with an area of 650 sq km (250 sq mi) with the population of 6,809,970 (2011) and further 7,749,334 living in the metropolitan area, making Hyderabad the fourth most populous city and the sixth most populous urban agglomeration in India. The city has acquired a national and international image since mid-1990s with the growth of information technology (IT) and efforts by the state government to promote the same. The decadal growth of urban population of Hyderabad remained between 4-5 lakh between 1901-1931. Sudden surplus of above 11 lakh with a decadal growth of 58.4 per cent and 52.5 per cent was registered between 1931-1951 respectively. This is attributed to the better administration, development of industries, business, transport and infrastructure. The minus growth between 1911-1921 was due to dreaded diseases of cholera, plague, influenza, malaria and other infectious diseases. Substantial growth of 9 lakh or 41.6 per cent between 1981 and 1991 is reasoned for heavy migration within the state. Political changes over attracted Andhra migrant employees and entrepreneurs to settle in Hyderabad. This exodus was disrupted between 1991-2001 due to Telangana struggle. Hyderabad grew to a tune of 222 per cent over five decades 1951-2001 with 25 lakh people being added to the city.

The population growth analysis for 1981-1991 showed a decadal increase of 57 per cent. The MCH area had 42 per cent. But the surrounding municipalities registered a very high increase of 155 per cent. This indicates that the urban sprawl of the city exceeded several fold in area and population. This urban trend signifies the potential of the city peripheries for attracting and sustaining a high population. The south and south west of the city is expanding tremendously. This is attributed to international airport at Shamshabad in the south and hi-tech city in the west.

Table 4.1

Hyderabad: Growth of Population

Year	*Population*	*Growth Rate (%)*
1881	354962	--
1891	415039	16.92
1901	448466	8.05
1911	502104	11.96
1921	405630	- 19.21
1931	466894	13.13
1941	739159	58.31
1951	110688	52.97
1961	1249151	10.48
1971	1796339	43.81
1981	2545836	41.72
1991	4344437	70.65
2001	5751780	32.39
2011	6809970	87.2

Source: Census of India Reports.

Table 4.2

Past, Present and Future Trend of Growth of Population

S.No.	Component	Area	Population in Lakh						Density of Population (sq km)		
			1981	1991	2001	2005	2011	2021	2001-2015	2011	2021
1	MCH	172	21.0	30.5	36.3	39.1	43.3	51.7	21048	25116	29977
2	Surrounding Municipalities	419	3.8	9.9	17.0	21.8	28.9	50.9	4102	6926	12151
3	Others	187	2.3	3.2	4.0	4.4	5.0	6.3	2147	2692	3391
4	HUA	778	27.1	43.6	57.5	65.4	77.2	108.9	7393	9923	13997
5	HUDA	1905	29.9	48.7	63.8	74.5	90.5	136.4	3351	4753	7162

Impact on Available Residential Space

A key feature of population growth is that most of the growth is occurring in the surrounding areas. The population growth in surrounding municipalities was phenomenal between 1981-1991 at 158 per cent as against 42 per cent in the MCH areas for the same period. Between 1991-2001, the population growth in surrounding municipalities stood at 72 per cent as against 19 per cent for the MCH area. Most of the growth has taken place away from the city core. The rates of growth of population experienced by HUA during 1991-2001 will continue in future though at a lesser rate. The projections indicate that the metropolitan area would house a population of 90.55 lakh in 2011 and 136.44 in 2021 including the population of the city.

HUA extends to over 778.17 sq km and is spread diagonally for 36 km from northwest to southeast and over 35 km in northeast-southwest direction. It is fifth fastest growing city in India and is currently home to over 6 million persons. Hyderabad Urban District came into existence in August 1978 consequent on the formation of a new district—Rangareddy—consisting of rural areas of Hyderabad. In 1995, HUDA was constituted with a jurisdiction of over 1864 km.[2]

Table 4.3

Trends in Growth of Constituents of HUA and HUDA Regions

Components of HUA/HUDA	Area (km^2)	Population (million)		
		1981	1991	2001
MCH	172.6	2.15	3.04	3.63
10 Municipalities	418.58	0.38	0.99	1.71
Secunderabad Cantonment	40.17	0.13	0.17	0.20
Osmania University and 13 Outgrowth, 5 Census Towns	146.82	0.09	0.15	0.19
Total of HUA region	778.17	2.76	4.36	5.75
Medchal, Gandipet, Panchayat Areas, Ghatkesar Census Town, etc.	1273.69	0.22	0.30	0.63
Total HUDA Region (incl. Secunderabad Cantonment)	1905.04	2.99	4.66	6.38

Table 4.4

Past, Present and Future Trend of Growth of Population

S.No.	Component	Area	Population (in lakh)						Density of Population (sq km)		
			1981	1991	2001	2005	2011	2021	2001-2005	2011	2021
1	MCH	172	21.0	30.5	36.3	39.1	43.3	51.7	21048	25116	29977
2	Surrounding Municipalities	419	3.8	9.9	17.0	21.8	28.9	50.9	4102	6926	12151
3	Others	187	2.3	3.2	4.0	4.4	5.0	6.3	2147	2692	3391
4	HUA	778	27.1	43.6	57.5	65.4	77.2	108.9	7393	9923	13997
5	HUDA	1905	29.9	48.7	63.8	74.5	90.5	136.4	3351	4753	7162

Table 4.5

Components of Population Growth (HUA)

Composition/ Year	Population Increase During			
	1981-1991 (in lakh)	Per cent of Total	1991-2001 (in lakh)	Per cent of Total
Natural Increase	8.7	59	9.0	70
Immigration	3.7	25	3.1	24
Jurisdictional Change	2.4	16	0.8	6
Total Change	14.8	100	12.9	100

Future Growth

The rates of growth of population experienced by HUA during 1991 and 2001 will continue in future though at a lesser rate. The projections indicate that the metropolitan area would house a population of 90.55 lakh in 2011 and 136.44 lakh in 2021 including the population of the city. Population growth characteristics of HUA are evaluated by looking at the past growth trends. The projections indicate that the city will have a population of about 4 million in 2006, 4.5 million 2011 and expected to cross the 5 million mark in 2021. The outer municipalities are growing at a faster pace than that of the Municipal Corporation. Accordingly, the forecasts for 2011 and 2021 indicate that the population is expected to be in the range of 29 lakh and 51 lakh respectively.

An analysis of composition of population growth indicates that natural growth contributed about 60 per cent while migration contributed 25 per cent in the year 1991. However, the contribution of natural growth has increased significantly in the year 2001 to 67 per cent. This indicates that the urban growth in HUA will take place even if measures are taken to control migration. Hence long-term and strategic planning is necessary to promote equitable economic growth and service delivery. Table 4.4 gives the details on composition of population growth. Most of the growth in HUA is occurring in the surrounding municipalities ranging from 56 per cent to 103 per cent, whereas the population growth in MCH area is only 19. 3 per cent. All municipalities have grown by more than 50 per cent compared to MCH and Secunderabad Cantonment.

Land Use for Sustainability

Hyderabad had an initial areal extent of 2.5 sq miles of walled city in 1591 AD. The Census of India was established in 1881 AD. It is noted that the city had 22 sq miles by then. In a period of three decades by 1911, the city area was doubled (50 sq miles); for over another two decades, it remained almost constant; during the year 1941, it raised to 78.54 sq miles; the next decade it was 83.3 sq miles. But the real change is noticed between 1961 and 1971 with the city having 172.60 sq miles—an increase of 89.3 sq miles—of

MCH area. This was the saturation and even in 2001, the area remained constant.

Land use in Hyderabad is of a mixed nature, given the multi-stage growth of the city and the scenic of several nuclei to give the amalgam that the present city is. As the city started expanding in a spatial sense, the third dimension was not to be left far behind and the skyline of the city started showing visible signs of an up-thwart. While earlier it was the central city location which witnessed these phenomena, very soon the periphery of the city also started showing perceptible signs of the third dimension and now it is very prominent in certain pockets and corridors on the periphery.

Table 4.6

Area Growth of Hyderabad (MCH) 1881-2001

Year	*Area (in sq miles)*	*% Increase*
1881	22	
1891	26	18.18
1901	26	0.00
1911	50.98	96.08
1921	51	0.04
1931	53	3.92
1941	78.54	48.19
1951	83.3	6.06
1961	85.08	2.14
1971	172.6	102.87
1981	172.6	0.00
1991	172.6	0.00
2001	172.6	0.00

Source: *Census India*, Population Tables, 1989.

The scenario in 2001 showed that about 44 per cent land was covered with residences, about 7 per cent increase over 1993. Commercial land use increased from 2 per cent in 1993 to 12 per cent in 2001, a spurt of 10 per cent. This shows the enormous commercialisation within MCH in a short span of time. While manufacturing decreased from 2.20 per cent to 1.8 per cent, public and semi-public land use increased from 8.8 per cent to 13.6 per cent a substantial increase (Table 4.4).

Land under graveyards, recreation, hillocks and vacant land category accounted for 33 per cent in 1993 but has come down to 16 per cent in 2001, the largest proportion under this category hillocks/ vacant land seems to have been built up and occupied. Further it is a matter of grave concern that the water bodies of MCH which made upto 12.6 per cent of the land use in 1993 declined by 7 per cent in seven years to show 5 per cent occupancy in 2001. A huge area in terms of road widening is acquired by MCH. Land use for transport and communication has increased from 5.1 per cent to 7.8 per cent.

Table 4.7

Hyderabad: Land Use/Land Cover (2002/2006)

Class	*Per cent Area (2002)*	*Area (in sq km)*	*Per cent Area (2006)*
Build up land	38.4	59.81	34.61
Vegetation	30.5	46.49	26.90
Water Bodies	3.2	8.03	4.65
Marshy Land	2.6	0	0.0
Open Scrub/spaces	7.2	23.63	13.79
Forests and gardens		34.64	20.05
Total Area	100.00	1726.0	100.00

Table 4.8

Land Use of Hyderabad

S. No.	*Category*	*1993*	*Per cent*	*2001*	*Per cent*
1.	Residential	63.38	36.72	75.2	43.56
2.	Commercial	3.00	1.74	20.6	11.94
3.	Manufacturing	3.76	2.20	3.07	1.78
4.	Public/Semi Public	15.07	8.78	23.48	13.62
5.	Utility including Graveyards	3.47	2.01		
6.	Recreation	5.35	3.10	28.14	16.30
7.	Agricultural Hillocks/Vacant	47.92	27.76		
8.	Water bodies	21.79	12.62	8.63	5.00
9.	Transport and Communication	8.86	5.13	13.48	7.80
	Total	172.60	100.00	172.60	100.00

Source: Vision 2020.

The city has densely built-up area at times interspersed by vegetation and open spaces which are more on the periphery. There are also quite a few water bodies, dotting the urban landscape, the prominent among them being lakes Hussain Sagar and Mir Alam and River Musi. Thick vegetation is more on the western part of the city. The 2006 statistics reveal that built-up area covers 59.81 sq km; followed closely by vegetation cover with 46.49 sq km. Forest and gardens come close third with 34.64 sq km, open spaces with 23.83 sq km area coverage. Last is the water body with a meager of 8.03 sq km.

It is evident that high density vegetative index is found on the western margins of the city towards Golconda, around Jubilee Hills, Miralam Tank, south-east parts of the city Santoshnagar, Midhani, DRDL, where mostly defence organisations are set up. The area under no vegetation cover is estimated to be 79.47 sq km area, followed by low density vegetation cover of 53.02 sq km area and high density vegetation cover of 40.11 sq km area. The extent of built-up area, 'other land' and vegetative cover increased by 355.6 sq km, 351.6 sq km and 1215.8 sq km respectively. However, area under water body and agriculture decreased by 248.5 sq km and 1474.5 sq km respectively.

A major change entitling shift of agricultural land to 'other land use' category is owing to rapid urbanisation in the region. Another reason which has fuelled this shift is the recurrent drought forcing farmers to opt for long fallow. This trend has affected the socio-economic condition in the region where large-scale migration of farming families to urban centres in search of livelihood, recurrent crop failure, high value of land, availability of credit facility, decreasing water resource, education, market economy, etc., have given rise to an unsustainable manner of economic development. In peri-urban area or rural belt, a number of fruit orchards and tree plantations have been established by large business houses that have brought land from small farmers and consolidated them into large holdings. These plantations and orchards account for the increase in vegetative cover in rural belt. Areas under water body have decreased as tank bottom and riverbeds are encroached for cultivation using conserved moisture. A few tank beds have been encroached upon to set up residential colonies.

Change during 1989-2006

Landsat TM image of two time period is taken to analyse the changes that have taken place over this time period. It is clear that 'maximum increase' has taken place in the entire city, more so towards the east interspersed with 'some change in increase'. Towards the west, the pockets of change have 'decreased', 'partially decreased' or remained unchanged between 1989-2006, for the obvious reasons that, this region being an undulating land covered with forest canopy does not give much scope for future increased activities. These pockets seem to have retained their inherent original cover with minimal changes. On the other hand, the built up cover seems to have expanded enormously with increasing pressure on land. The intensity of built up activity is directed both horizontally as well as vertically. The new trend of multistoried buildings due to ever-increasing demand for housing has given rise to lesser floor space per person. More area under 'some increase' in change is observed towards west and south, mostly interspersed with decreased change.

Table 4.9

Hyderabad City: Change Detection of LULC (1989-2006)

Class Name	*Area (sq km)*	*Per cent*
Decreased	1.47	0.85
Increased	105.57	61.20
Some Decreased	3.58	2.07
Some Increased	60.42	43.97
Unchanged	1.58	0.91
Total	172.6	100.00

It is clear that a maximum of 105.75 sq km (61.20%) of the city land area has 'increased' with regard to built-up activity, followed closely by 'some increase' with 60.42 sq km (34.97% of area). In Contrast, 'some decrease', 'decreased' and 'unchanged' has area well below 5 sq km, almost negligible in comparison to the area under 'increase'. This change detection clearly brings out the urban growth process between 1989-2006, with accompanying land use/ land cover change in built-up activity.

Table 4.10

Hyderabad City: Land Use Classification

Type	*Total area (km²)*		
	1975	*1989*	*2003*
Marshy land	76.608171	29.6325045	27.7513335
Deep water	9.415602	52.23985875	27.79763175
Scrub	427.616974	300.616974	300.616974
Shallow water	25.384437	44.417079	14.6838555
Built up	317.807433	475.6371963	745.8006398
Bare soil	392.368734	185.3075273	7.566921
Barren land	579.36168	381.1548105	551.6111588
Vegetation	78.908463	147.3023498	68.91129
Type	76.608171	29.6325045	27.7513335

On the contrary the real estate investment in HUA had taken place in the outer areas resulting in the development of scattered built-up pockets leaving behind vacant spaces in between. As the city sprawl increased in dispersed low density pattern, the urban expansion devoured the agricultural lands and water bodies. In Hyderabad, the residential land use of fringe area has increased from 147.03 km^2 (8.69%) in 1993 to 214.17 (12.66%) in 2000. On the contrary the agricultural and vacant land use has declined from 1209.17 km^2 (71.45%) in 1993 to 1117.73 km^2 (66.05%) in 2000.

The area under vegetative cover increases outward from city centre making quality of life poorer at the centre. In the city centre available green cover per person is meagre of 5.9 square metres (sq mt), it increases 15.5 sq. mt in urban belt, 138.2 sq mt in sub urban and over 1117.3 sq mt in rural areas.

In order to achieve sustainable urban planning and to check haphazard development, it is necessary that authorities associated with the urban development generate such planning models so that every bit of the available land is used in most rational and optimal way. This requires the present and past land use/land cover information of the area and pattern of changes with respect to urban settlements and other local resources.

The scenario in 2001 showed that about 44 per cent land was covered with residences, about 7 per cent increase over 1993. Commercial land use increased from 2 per cent in 1993 to 12 per cent in 2001, a spurt of 10 per cent. This shows the enormous commercialisation within MCH in a short span of time. While manufacturing decreased from 2.20 per cent to 1.8 per cent, public and semi-public land a substantial increase.

Impact on Land Use-Land Cover

Although built-up area is predominant only in the city centre and accounts for 7.3 per cent of the entire region, its impact on the natural resource base in the region is massive. Urbanisation has alternated the land use-land cover in the region with agricultural land being converted into settlements, roads, railways and commercial complexes.

This trend indicates an increase in urban sprawl in rural belt and rural area at the cost of agriculture which is the main source of livelihood of rural poor. Such a trend is unsustainable as the rural population is ill-equipped to the changed circumstances. With loss of agricultural land, they are forced to migrate to urban areas to look for jobs and slums are springing up in urban areas, thus degrading the environment in urban centres.

Table 4.11

Per cent Share of Various Land Use-Land Cover in the Region

Zones (Distance from Centre)	*Per cent Share in Zones*					
	Built-up Area	*Urban Vegetation*	*Agriculture Land*	*Road (km)*	*Other Land*	*Water Body*
Rural Area (25-50 km)	4.5	21.5	51.4	5182. 9	22.1	0.4
Rural Belt (10-25 km)	9.9	16.7	53.6	2187.7	17.9	1.9
Urban Belt (5-10 km)	34.8	22.6	31.6	840. 2	9.4	1.6
City Centre (0-5 km)	75. 9	15.9	0.0	748.1	2.5	5.7

Impact on Environment

Due to high concentration of urban population in Hyderabad, agricultural activity has suffered great losses by way of shrinking areal extent and falling productivity. For obvious reasons the percentage share of population in agriculture in the city centre of Hyderabad is a meagre 0.2 per cent while the peripheral regions of Mahabubnagar, Medak and Nalgonda have over 23 per cent of its adult population engaged in agriculture.

Due to rapid pace of urbanisation, agriculture—the mainstay of AP state economy—has been relegated to the background. Prime agricultural lands around the city are converted into residential, commercial or industrial use. Apart from this large tracts are left fallow owing to various reasons like lack of adequate irrigation sources, failure of tube wells owing to fall in groundwater level, erratic rainfall, migration of land owners to urban centres, ineffective laws of tenancy, lack of input with farmers owing to poverty and shortage of labour during critical phases in cropping season, etc. As a consequence, agricultural workers in Hyderabad Urban District constitute less than 0.2 per cent of total workforce. On the other hand, agriculture is an important economic activity and source of livelihood in rural areas in Medak (30%), Mahabubnagar (42%) and in Nalgonda district (23.5%), while Rangareddy with over 16 per cent workers depend on agriculture for a livelihood.

Impact on Land Use

Although built-up area is predominant only in the city centre and accounts for 7.3 per cent of the entire region, its impact on the natural resource base in the region is massive. Urbanisation has alternated the land use-land cover in the region with agricultural land being converted into settlements, roads, railways and commercial complexes.

Urban Sprawl and Sustainable Development

Analysis of the data suggests a clear increase in residential, commercial, industrial and transportation zones in the urban area. In the non-urban area, there is a clear reduction in agriculture area and also in vacant land suggesting the increased intensity with urbanisation activities. The results suggest that there is substantial variation in the patterns of urban sprawl among the different zones of the study area corresponding to residential, industrial, commercial, sensitive and mixed zones.

Analysis of growth in built-up area over a 30-year period indicates highest maximum increase in the 10-25 km rural belt by 428.8 per cent. In rural area (25-50 km belt), area under settlement increased by 305.5 per cent during the corresponding period, while in 5-10 km urban belt, this increase was only 37.8 per cent and in the city centre this growth was only 1.02 per cent. Overall growth in extent of built-up area in the study area was 167.89 per cent.

In order to sustain development, the supply and quality of major consumables and inputs to our daily lives and economic production—such as air, water, energy, food, raw materials, land and the natural environment—needs to be taken care of. Land is essential because our food and raw materials originate from them and is a habitat for flora and fauna. Similar to other resources, it is a scarce commodity. Any disturbance to this resource by way of change in land use, e.g., conversion of forestland, agricultural land into built-up, is irreversible. The use of land unsuitable for development may be unsustainable for the natural environment as well as to the humans.

In peri-urban area or rural belt, a number of fruit orchards and tree plantations have been established by large business houses that have brought land from small farmers and consolidated them into large holdings. These plantations and orchards account for the increase in vegetative cover in rural belt. Area under water body has decreased as tank bottom and riverbeds are encroached for cultivation using conserved moisture. A few tank beds have been encroached upon to set up residential colonies—a most undesirable trend, compounding the problem of water shortage in the study area.

For sustainable development of the city; to ensure adequate power throughout the year without the need for regular power cuts, and with a potential to maintain comfort conditions for nearly 80 per cent of the time using passive solar features or low energy systems, energy efficiency of building in Hyderabad has to be improved immensely. Especially, reducing the energy intensity of commercial buildings will be very important, particularly as commercial area within Hyderabad continues to increase.

Summary and Conclusions

Hyderabad is one of the leading metropolitan cities of India. The city has under gone considerable socio-cultural change in the past. Today also it is undergoing rapid and dynamic changes due to accelerated phase of industrialisation and urbanisation. These changes had a deep impact on the pattern and developmental trends in the city.

To address the problems of Hyderabad city is to support urban reforms to improve city management and upgrade municipal services through provision of physical infrastructure with particular focus on reducing urban poverty. Improving urban land planning and management assessment includes collection of the social and environmental data from the secondary sources, social and environmental assessment for an in- depth assessment of the institutional capacity.

The growth of the city has occurred at a much faster rate in the peripheries in the last couple of decades than in the metropolitan core. Accordingly, a number of new activities are emerging in the periphery with the active participation of both the state and the private sector. These trends have significant implications on land use, urban planning and policies in the coming years.

The population growth analysis for 1981-1991 showed a decadal increase of 57 per cent. But the surrounding municipalities registered a very high increase of 155 per cent. This indicates that the urban sprawl of the city exceeded several fold in area and population. This urban trend signifies the potential of the city peripheries for attracting and sustaining a high population. This is attributed to the international airport at Shamshabad in the south and the hi-tech city in the West.

The scenario showed that about 44 per cent land was covered with residences, about 7 per cent increase over 1993. Commercial land use increased from 2 per cent in 1993 to 12 per cent in 2001, a spurt of 10 per cent. This shows the enormous commercialisation within MCH in a short span of time. While manufacturing decreased from 2.20 per cent to 1.8 per cent, public and semi-public land use increased from 8.8 per cent to 13.6 per cent a substantial increase. Land under graveyards, recreation, hillocks and vacant land category accounted for 33 per cent in 1993 but has come down to 16 per cent in 2001, the largest proportion under the category hillocks/ vacant land seems to have been built-up and occupied. Further it is a matter of grave concern that the water bodies of MCH which made upto 12.6 per cent of the land use in 1993 declined by 7.0 per cent in seven years to show 5.0 per cent occupancy in 2001. A huge area in terms of road widening is acquired by MCH. Land use for transport and communication has increased from 5.1 per cent to 7.8 per cent.

The 2006 statistics reveal that built-up area covers 59.81 sq km; followed closely by vegetation cover with 46.49 sq km. Forest and gardens come close third with 34.64 sq km, open spaces with 23.83 sq km area coverage. Last is the water body with a meagre 8.03 sq km. The area under no vegetation Cover to be 79.47 sq km area, followed by low density vegetation cover of 53.02 sq km area and high density vegetation cover of 40.11 sq km area.

Analysis of growth in built-up area over a 30-year period indicates highest maximum increase in the 10-25 km rural belt by 428.8 per cent. In rural area, (25-50 km belt) area under settlement increased by 305.5 per cent during the corresponding period while in 5-10 km urban belt, this increase was only 37.8 per cent and in the city centre this growth was only 1.02 per cent. Overall growth in extent of built-up area in the study area was 167.89 per cent.

This trend indicates an increase in urban sprawl in rural belt and rural area at the cost of agriculture which is the main source of livelihood of rural poor. Such a trend is unsustainable as the rural population is ill-equipped to the changed circumstances. With loss of agricultural land, they are forced to migrate to urban areas to look for jobs and slums are springing up in urban areas, thus degrading the environment in urban centres. Spatial analysis indicates that urban growth is eastward along the highways. Dispersed growth is observed in north-west sector around Patancheru. Growth is however, constrained in south-west sector owing to presence of hillocks and an undulating topography.

Recommendations

Hyderabad has transformed from a conventional city. The city proudly boasts of a strong economy fuelled by its industries—keeping intact the social, cultural and educational strongholds. The civic infrastructure growth and service standards however fall short of the actual requirements. The Jawaharlal Nehru National Urban Renewal Mission (JNNURM) has come as a big opportunity to enhance the city's infrastructure image, quality of service and standard of living.

It has brought with itself immense responsibility on the people and the city as a whole. While reforms need to be pursued vigorously for improved service delivery, to really exploit the massive inflow of funds coming into the city of Hyderabad, very few of the infrastructure projects planned have been delivered fully in the two years. In a more optimistic vein, progress should not go unnoticed and has to be appreciated.

JNNURM addresses much of the infrastructure and service delivery needs. But monitoring mechanisms should be strengthened and decentralised to confirm that projects are taken up and executed responsibly. There is great scope for improvement. It is the responsibility of the GHMC to undertake improvements and bring about clean execution of the mission. JNNURM is only a stepping stone towards attaining the vision envisaged.

Relocation of certain amenities and facilities should be done to relieve traffic and maintain balanced location and distribution. Expand arterial roads along with the main roads, plan for Multi Model Transport System (MMTS) along with roads wherever possible and provide multilevel parking using open spaces to free congestion on roads. Flyovers/foot over bridge up to a certain limit can reduce congestion which in return can reduce pollution. Smooth flow of traffic without abstracters could reduce air pollution.

The concept of public-private participation in any development plan would lead to sustainable growth and development of the city. Encourage social forestry, gardens, greenery in and around Hyderabad. Encourage private investment for urban infrastructure/ amenities and facilities and collect user charges whenever necessary.

In order to maintain the city's identity, we need to seek solutions to conserve our unique ecology and geological creations while allowing for growth. From what we can see, we can clearly see that the development is far from being sustainable. While the government is keen on moving in the right direction in this matter, there are other players who can also make a substantial difference. Hence, let us now look at improving environment.

Considering the current growth patterns, it advocates clustering of developed areas for a compact development instead of a sparse development, which is currently being witnessed. It promotes a corridor development but with adequate precautions in the form of controls, access and services. This underlines the need for an environmentally sustainable planning through appropriate planning for services and industrial areas. The future expansions are to be accommodated within or around the current developments in conjunction with the development of the transport corridors.

HUA has witnessed many land use-land cover changes. Most of the trends are unsustainable. This is compounded by weak administration because of which the pressure on existing resources would rise to a breaking point. Poor infrastructure facilities and employment opportunities in rural area would encourage continued migration towards urban agglomeration and in that process create bottlenecks in urban planning. More land use changes are inevitable in the long run but it is necessary to safeguard sustainable trend in economic development. One solution to this problem could be the development of satellite townships in order to decongest the city centre.

References

Alam, S.M. (1965). *Hyderabad-Secunderabad: A Study in Urban Geography.* Bombay: Allied Publishers Pvt. Ltd.

———. (1984). *Vulnerability and Resilience of Cities: The Case of Hyderabad.* UNESCO Pilot Project Report. Hyderabad: Centre for Area Studies, Osmania University.

Alam, S.M. and V.V. Pokshishevsky (eds) (1976). *Urbanisation in Developing Countries.* Hyderabad: Osmania University.

Alam, S.M and W. Khan (1972). *Metropolitan Hyderabad and its Regions: A Strategy for Development.* Mumbai: Asia Publishing House.

Boss, Ashish (1973). *Studies in India's Urbanisation: 1970-71.* New Delhi: Tata McGraw Hill Publishing Company limited.

CDA (2001). *Master Plan for Cyberabad Development Authority Area.* Hyderabad.

Census of India (2001). "Series 29 A.P. Provisional Population Totals, Paper 2 of 2001", *Rural and Urban Distribution of Population.* New Delhi: Controller of Publications.

Chandra Shekar, A. (1961). Quoted in *History of Hyderabad: A. Press Reporters Guide.* Hyderabad.

Gopi, K.N. (1978). *Process of Urban Fringe Development.* New Delhi: Public Concept Publishing Company.

———. (1980). *Urban Growth and Industrial Locations.* New Delhi: Oxford and IBH.

Ramchandriah. C. (2003). "Information Technology and Social Development", in C.H. Hanumantha Rao and S. Mahendra Dev (eds.) (2003), *Andhra Pradesh Development: Economic Reforms and Challenges Ahead.* Hyderabad: Centre of Economic and Social Studies.

Ramachandraiah, C. and Prasad Sheela (2004). "Impact of Urban Growth on Water Bodies The Case of Hyderabad", *Working Paper* No. 60. September. Hyderabad: Centre of Economic and Social Studies.

Ramachandran Kausalya (2001). "Assessment of Contamination of Natural Resource using GIS and Remote Sensing: A Study on Hyderabad Region", *GIS India* 10(1).

Ramachandran Kausalya and H.P. Singh (2001). *'Sectoral Shift in Land Use under Changed Economic Scenario: A Case Study of Hyderabad Region using IRS-1D LISS-III Data.* ICORG 2-5 February. Hyderabad: JNTU.

5

V.G. RANGANATH

Precautionary Principle

An Environmental Perspective

> We should be on our guard not to overestimate science and scientific methods when it is a question of human problems; and we should not assume that experts are the only ones who have a right to express themselves on questions affecting the organisation of society.
>
> —Albert Einstein

Introduction

'Precautionary principle' is one of the most contentious principles in contemporary International legal developments. The nature and scope of its application has varied widely according to the context and sector within which it has been applied.

Precautionary principle plays a significant role in determining whether developmental process is sustainable or not. Precautionary principle underlies sustainable development which requires that the developmental activity must be stopped and prevented if it causes serious and irreversible environmental damage. The emergence of precautionary principle marks a shift in the international environmental jurisprudence—a shift from assimilative capacity principle to precautionary principle.

The precautionary principle is enshrined as one of the principles of the Rio Declaration 1992 from United Nations Conference on Environment and Development, also known as Agenda 21. The declaration stated: '

In order to protect the environment, the precautionary approach shall be widely applied by States according to their capabilities. Where there are threats of serious or irreversible damage, lack of full scientific certainty shall not be used as a reason for postponing cost effective measures to prevent environmental degradation.

Precautionary Principle: Judicial Activism

In Vellore Citizens Welfare Forum *versus* Union of India, where Supreme Court referred the Brundtland Report and other international documents in addition to Articles 21, 48A and 51A(g) of the Constitution of India. And also taken into account the legislative mandate "to protect and improve the environment" as found in enactments like the Water (Prevention and Control of Pollution) Act, 1974, the Air (Prevention and Control of Pollution) Act, 1981, and the Environment (Protection) Act, 1986.

In M.C. Mehta *versus* Union of India (Taj Trapezium Case), the Supreme Court was dealing with the problem of protecting the 'Taj Mahal' from the pollution of nearby industries. The court applied the precautionary principle' as explained by it in Vellore case and observed: The environmental measures must anticipate, prevent and attack the causes of environmental degradation. The onus of proof is on an industry to show that its operation with the aid of coke/coal is environmentally benign. It is rather, proved beyond doubt that the emissions generated by the use of coke/coal by the industries in Taj Trapezium are the main polluters of the ambient air. The court ordered the industries to change over to the natural gas as an industrial-fuel or stop functioning with the aid of coke/coal in the Taj trapezium and relocate themselves as per the directions of the court.

In Narmada Bachao Andolan *versus* Union of India, precautionary principle came to be considered by the majority of judges. The court also took the view that the doctrine is to be employed only in cases of pollution when its impact is uncertain and non-negligible. In Andhra Pradesh (AP) Pollution Control Board *versus* M.V. Nayudu, the Supreme Court has reiterated its earlier stand on the precautionary principle and demanded that the burden of proof should rest with the person/entity proposing the activities which may have harmful effects on the environment and/or human beings.

General Food Law: Precautionary Principle

Article 7 Regulation EC/178/2002 formally establishes the precautionary principle as an option open to risk managers when decisions have to be made to protect health but scientific information concerning the risk is inconclusive or incomplete in some way.

The precautionary principle is relevant in those circumstances where risk managers have identified that there are reasonable grounds for concern that an unacceptable level of risk to health exists but the supporting information and data may not be sufficiently complete to enable a comprehensive risk assessment to be made. When faced with these specific circumstances, decision makers or risk managers, may take measures or other actions to protect health based on the precautionary principle while seeking more complete scientific and other data. Such measures have to comply with the normal principles of non-discrimination and proportionality and should be considered as provisional until such time that more comprehensive information concerning the risk can be gathered and analysed.

The technology of GE (genetic engineering) is a fit case for the application of the precautionary principle. The precautionary principle necessitates that if there are reasonable scientific grounds for believing that a new process or product may not be safe, it should not be introduced until we have convincing evidence of reasonable certainty of no harm. The principle can also be applied to existing technologies when new evidence appears, suggesting that they are more dangerous than what society had previously expected. Then, it requires that the research to better assess the risk and that in the meantime the technology should not be used and expanded. If the dangers are considered serious enough, then the principle may require us to withdraw the products or impose a ban or a moratorium on further use.

The Apex Court of India in AP Pollution Control Board *versus* M.V. Nayudu held that precautionary principle is applicable in India. The principle mandates that when a new technology or process can cause serious and irreversible harm to human health and the environment, precautionary measures should be taken even if some cause and effect relationships are not fully established scientifically. In this context, the proponent of the novel and uncertain activity rather than the public should bear the burden of proof. As, if one is embarking on something new, one should go ahead only and until one is reasonably convinced that it is safe. Pushing forward with untested, inadequately researched technologies, and insisting that it is for society to prove conclusively that they are harmful before they can be stopped, is self-defeating and extremely dangerous.

Nowadays, it become the common practice to introduce the food in new terms by applying the biotechnology or genetically modified/engineered. There should be adequate scientific evidence in terms of research carried out in various parts of the world, to at least doubt the safety of the process of GE. The author suggests that in the interest of the public, the precautionary principle be employed and a moratorium be ordered on any further release till there is reasonable certainty of the safety of such releases, through independent scientific studies. Such studies would in the normal course take a number of years, akin to the scientific studies mandatory before the approvals given for pharmaceuticals. Such moratorium should also be accompanied with isolation, destruction and recall of the GMOs (genetically modified organisms) that have already been released, to the extent that is possible. It should be instructive that the proponents of technology also refuse to accept liability. The implication, therefore, is that if the technologies turn out to be hazardous, as in many cases they have, the rest of society is left to pay the penalty. In a major setback to the proponents of genetically modified technology in farm crops, the Parliamentary Committee on Agriculture on 9 August 2012 asked the government to stop all field trials and sought a bar on GM (genetically modified) food crops (such as Bt brinjal).

India's Obligations under International Agreements in Outlook of Genetically Modified Organisms

The Food Safety stand has to be particularly focused mainly on experimentation and allowing of GMOs and inserting into food crops. The Cartagena Protocol on Biosafety for the Convention on Biodiversity[1] was adopted in 2002 and came into force on 11 September 2003. It is a binding agreement on biosafety. India is a signatory and is bound to implement its provisions. Article 1 of the Protocol lays down the objective to contribute to ensure adequate levels of protection in the field of safe transfer, handling and use of living modified organisms (LMOs or GMOs) that may have adverse effects on the conservation and sustainable use of biodiversity, taking into account the risks of human health. Article 2(2) stipulates that parties must ensure that the development, handling, transport, use, transfer, and release of LMOs are undertaken in a manner that prevents or reduces risk to biodiversity. Further, Article 10(7) (Precautionary Principle) necessitates that the lack of scientific certainty due to sufficient relevant scientific information and knowledge regarding the extent of potential adverse effects shall not prevent the contracting party from taking a decision, as appropriate, in order to avoid/minimise potential adverse effects. According to Para 9(h) of the Annexure III of the Protocol, the parties are bound to consider information on the location, geographical, climatic and ecological characteristics, including relevant information on biodiversity and centres of origin of the likely potential receiving environments. Annexure 3 of the Protocol includes, *inter alia*, the general principles of risk assessment: (i) risk assessment should be carried out in a scientifically sound and transparent manner, and can take into account expert advice as well as guidelines developed by relevant international organisations. According to Article 27 of the Protocol parties may account for socio-economic considerations arising from impact of LMOs on

1. The full name of the Bio-safety Protocol is 'The Cartagena Protocol on Bio-safety to the Convention on Biological Diversity'. Cartagena is the name of the city in Colombia where the Bio-safety protocol was originally scheduled to be concluded and adopted in February 1999. However due to a number of out standing issues, the protocol was finalised and adopted a year later on 29 January 2000 in Montreal, Canada, by more than 130 countries. The protocol entered into force on 11 September 2003. As on date, 99 countries ratified the protocol. India has also ratified the Cartagena Protocol on Bio-safety on 23 January 2003.

biodiversity conservation and sustainable use, especially with regard to value of biodiversity to indigenous and local communities. According to Article 23(iii) of the Protocol, parties shall consult the public in the decision-making process regarding LMOs; while Article (iv) mandates that parties shall make decisions available to the public, but respecting confidential information. Further, Article 21(7) prescribes that the information about a summary of the risk assessment cannot be made confidential.

The UN Convention on Biological Diversity (CBD), 1992, *inter alia* requires that the contracting parties to domestically regulate or manage the risks associated with the use and release of LMOs resulting from biotechnology and which are likely to have adverse environmental impacts that could affect the conservation and sustainable use of biodiversity, and the risk to human health, introduce appropriate procedure to require impact assessment of proposed projects likely to have significant adverse effects on biodiversity and to allow public participation in the procedure where appropriate (Article 14). Further, Article 19(3) of the CBD had urged parties to consider the need for and modalities of a protocol setting out appropriate procedures in the field of the safe transfer, handling and use of any LMOs that may have adverse effect on the conservation and sustainable use biodiversity.

Despite the eroding effects of the insidious relationships between research, regulation and industry, strong public pressure has, nevertheless, forced numerous regions of Europe and other parts of the world to declare themselves GM free. Over 58 countries have enacted biosafety laws to restrict import and commercialisation of GM products and/or require labelling of food containing GM ingredients. More than 100 regions and 3500 sub-regions in Europe, the most important market for such products, have declared themselves GMO-free, and have demanded new European law to protect them from GM contamination.

Technical Expert Committee Report: Centre's Representation Before Supreme Court of India

The centre on 8 November 2012, informed the Supreme Court that the recommendations of the Technical Expert Committee

(TEC) seeking a 10-year moratorium on field trials on GM crops will be highly detrimental and will not be in national interest. "Based on current overall status of food safety evaluation of Bt. Transgenics, including the data on Bt cotton and Bt Brinjal examined by the TEC, and in accordance with the precautionary principle, the TEC recommends a 10-year moratorium on field trials of Bt. Transgenics in all food crops," the TEC said in its interim report submitted to the Supreme Court. "Another factor is the possibility of contamination of non-GM food by GM food." Appearing for the centre, Attorney-General contended that field trials should be allowed to go on as the demand for food for the growing population could be met only through the GM crops. Counsel for various parties also opposed the recommendations. Taking note of the objections against the report, the Supreme Court refused to stay the field trials and directed that all the objections be placed before the TEC for fresh consideration and asked the committee to submit its report in six weeks. The bench was hearing a petition filed by Aruna Rodrigues for a direction to ban such field trials. The committee was constituted by the court to give its recommendations to the government. The centre said adoption of the interim recommendations would bring to a halt the process of testing/assessment of the safety of GM crops.

Bt Cotton to Bt Brinjal

The most common insect-tolerant crops worldwide are Bt cotton and Bt corn, which are made by creating a cotton or corn plant with a gene drawn from a soil bacterium called *Bacillus thurengiensis* (it is the name of this bacterium shortened into Bt) that appears in the name of the GE crops, and the actual gene inserted in most Bt cotton hybrids in India is Cry1Ac. A Bt cotton plant is supposed to produce its own toxin from this *Bt* gene inserted into it to kill the American bollworm and the spotted bollworm. Bt cotton proponents say that it is also effective against the pink bollworm. The Indian experience of GM crops began as back in 2002 when three hybrid varieties of Bt cotton were grown.

In 2002, India allowed farmers to cultivate Bt cotton—the only GM crop commercially grown in India till date. In 2003, cotton farmers in the southern state of AP, whose government had aggressively promoted GM technology, suffered severe agricultural and financial losses and many including entire families committed suicide.

There was a huge uproar in the AP Legislative Assembly with the then government, then in the opposition, demanding compensation for Bt cotton farmers. Based on their own survey, on 10 March 2003, the then Minister for Agriculture in the Government of AP made a statement in the Assembly admitting that the performance of Bt cotton has been less than satisfactory. He said that "overall information is that the farmers have not experienced very positive and encouraging results" and that therefore, the farmers need to be compensated. To this, MMB (Monsanto-Mahyco Biotech) said that it will compensate farmers only for failure to germinate and for absence of genetic purity promised by the company and not for yield losses. This response from the company was not questioned by the government and a bad precedent was set, as similar results from 2004-05 would show. The AP Government did not make any changes to its 'MoU system' to prevent such responses from seed companies. The fiasco went unchallenged, while the first GE/Bt cotton farmer in the country committed suicide in AP in 2003 after the very first season of the commercialisation of Bt cotton. In the year of 2004, again, the AP Government had to officially order M/s Mahyco to pay compensation to the tune of ₹4 crore to Bt cotton farmers who lost their crop, which the company refused to. The losses were estimated only in a small part of Warangal district, whereas reports of failure emerged from several districts of the state. Due to Bt cotton cultivation, there were certain adverse effects on human health and the inexplicable phenomenon of toxicity in animals after grazing on Bt cotton fields. The Bt cotton experience shows that there has never been a more deserving case for applying the precautionary principle in the case of any technology. A toxin present in Bt cotton is so harmful. If people think that the problem caused by Bt cotton will confined to farmers and animals only, people need to think it again because that the gene that is used in Bt cotton is the same which is now being used in Bt brinjal. Naturally the people are outraged and they are out of street to register their protests.

Bt Brinjal Commercial Cultivation Put on Hold

Giving into intense opposition from NGOs (non-governmental organisations) and several states, the government today put on hold commercial cultivation of GM brinjal citing lack of clear consensus within the scientific community. The then environment minister, Jairam Ramesh announced the decision to put a moratorium on the release of Bt brinjal till such time independent scientific studies establish, to the satisfaction of both the public and professionals, the safety of the product from long-term view on impact on human health. The minister also made public a copy of his exhaustive report which relies on inputs received from stakeholders from all across the spectrum of scientists, civil society, academics, chief ministers of various states and others concerned.

Following is an excerpt from the statement issued on 9 February 2010:

> Based on all the information presented in the preceding paragraphs and when there is no clear consensus within the scientific community itself, when there is so much opposition from the state governments, when responsible civil society organisations and eminent scientists have raised many serious questions that have not been answered satisfactorily, when the public sentiment is negative and when Bt brinjal will be the very first genetically-modified vegetable to be introduced anywhere in the world and when there is no over-riding urgency to introduce it here, it is my duty to adopt a cautious, precautionary principle-based approach and impose a moratorium on the release of Bt brinjal, till such time independent scientific studies establish, to the satisfaction of both the public and professionals, the safety of the product from the point of view of its long-term impact on human health and environment, including the rich genetic wealth existing in brinjal in our country. A moratorium implies rejection of this particular case of release for the time being; it does not, in any way, mean conditional acceptance. This should be clearly understood. I am not getting into any time frame... The moratorium will continue for as long as it is needed to establish public trust and confidence.

Minister made it clear that the moratorium was only confined to Bt brinjal and the decision did not cover the larger issue of GE and biotechnology in agriculture.

Suggestions

1. There should be a proper system which will ensure that potentially hazardous GMOs are put through an adequate and proper safety tests by independent scientific agencies and whose results put to critique by independent experts so that the precautionary principle is adhered to before such organisms are released into environment.
2. GM crops should be introduced in our country only after proven tests and experiments.
3. The regulation and risk assessment of GM crops must address two questions. First, what is the degree of enforcement and commitment to keeping India safe? And second, what is required to regulate this hazardous technology before the environmental release of any GMO? The need for extreme caution arises from the fact that once released into the environment, these organisms cannot be recalled.
4. There are no liability laws in this country. For example, if a farm is growing Bt brinjal and the field beside it is producing organic brinjal, then who will take responsibility of mutation of the natural brinjal? The whole approach is quite absurd, unscientific and unplanned. It is no longer in question that GM technology is unpredictable and imprecise, that too when released in an open environment situation.
5. People must know what they are eating and to have a choice in this matter. The reality is that they are eating GM all the time without our knowledge.
6. The author hopes that genetic modification or genetic engineering is not the solution to the food security problem.

Instead, the government can enhance more funding for agriculture that emphasises solutions to the problems faced by poor farmers.

7. Recognising that safety assessment of a GM crop is the most important step, rules and procedures as per scientifically validated international norms should be in place for each stage of research and development.

Conclusion

The precautionary principle is relatable to risk assessment and environmental impact assessment. Broadly, it postulates that decisions that may have an impact on the environment need to allow for and recognise conditions of uncertainty, particularly with respect to the possible environmental consequences of those decisions. Under the circumstances, it is essential to take preventive action or avoid effects, which may be damaging even if this cannot be proven. In short, the 'precautionary principle' is a notion which supports taking protective action before there is complete scientific proof of a risk, that is, action should not be delayed simply because full scientific information is lacking. Previously, precautionary principle is being used only on the matter of environmental pollution but today is being used to deal with the wildlife protection, biodiversity conservation, matters related to climate change, protection of shrimps, etc. In May 2000, more than 69 eminent scientists of the world from 79 countries released an open statement about the hazards that the release of GMOs poses to biodiversity, food safety, and therefore to human and animal health, and they demanded a moratorium on the environmental release of such GMOs in accordance with the precautionary principles. This statement was issued on the eve of UN CBD Conference in Nairobi in May 2000.

References

Centre for Sustainable Agriculture (CSA) (2005). *The Story of Bt cotton in Andhra Pradesh: Erratic Processes and Results*. February. Available at: *http://www.grain.org/research_files/Story_of_Bt_Cotton.pdf* (accessed on 3 February 2014).

Gargi, Parsai (n.d.). ''Bar GM Food Crops, Says Parliamentary Panel'', *The Hindu*. Available at: *http://www.thehindu.com/news/national/article3747247.ece* (accessed on 3 February 2014).

Kuruganti, Kavitha and G.V. Ramanjaneyulu (2008). *Genetic Engineering in Indian Agriculture: An Introductory Hand Book*. Centre for Sustainable Agriculture, Secunderabad, available at *http://www.indiagminfo.org/updates/GE%20Handbook-CSA-march2008.pdf* (accessed on 3 February 2014).

Mukherjee, Debashree (2008). *Genetically Modified Crops-Issues and Perspectives*. Hyderabad: The ICFAI University Press.

Spardha (n.d.). "India says No to Bt Brinjal". Available at: *http://spardha.wordpress.com/2010/02/09/india-says-no-to-Bt brinjal/* (accessed on 3 February 2014).

United Nations General Assembly, *http://www.un.org/documents/ga/conf151/aconf15126-1annex1.htm*

6

R. REJANI, K.V. RAO, K.G.R. PRATHYUSHA, J.V.N.S. PRASAD, M. OSMAN, G.R. CHARY, PUSPANJALI, JOSILY SAMUEL and V.U.M. RAO

Spatial and Temporal Variation of Rainfall and AET in Low, Medium and High Rainfall Districts in Telengana Region of Andhra Pradesh

Introduction

Changing precipitation and its impact on surface water resources is one of the important problems faced by the world today. Many studies indicated significant changes in rainfall patterns both in global and regional scales. Long-term trends of Indian monsoon rainfall for the country as a whole and at regional scale have been studied by several researchers (Kumar et al. 2010). Guhathakurta (2005) implemented an Artificial Neural Network method in predicting monsoon-rainfall over Kerala, a southern state of India. Goswami et al. (2006) used high resolution gridded daily rainfall data and found that there are significant rising trends in the frequency and the magnitude of extreme rain events over central India during the monsoon season. The study also indicated significant decreasing trend in the frequency of moderate events during the same period, leading to no significant variation in the mean rainfall. Rainfall has an important role in the Indian agriculture since major portion of the agriculture area is under rainfed conditions and any variability in rainfall can significantly influence the sustainable management of water resources, agriculture and ecosystems. Aggarwal (2007) studied the effects of climate change and its implications for Indian agriculture. Rajeevan et al. (2008) analysed the variability and trends of extreme rainfall events over India using 104 years of gridded daily rainfall data.

Large variations in rainfall and its distribution can cause significant change in land use patterns. LULC (land use-land cover) changes have a large impact on climate, at local and regional levels due to modifications in the carbon cycle, ET (evapotranspiration) patterns and hydrological cycle. ET is the second largest component (after precipitation) of the terrestrial water cycle at the global scale, since ET returns more than 60 per cent of precipitation on land back to the atmosphere and thereby conveys an important constraint on water availability at the land surface (Kumar et al. 2011). Global climate change will affect precipitation and ET, and hence influence the renewable freshwater resources. Accurate estimation of ET not only meets the growing competition for the limited water supplies and the need to reduce the cost of irrigation projects, but is also essential to project potential changes in the global hydrological cycle under different climate change scenarios (Teuling et al. 2009). The actual ET is the real amount of water consumed by the crops, soil and vegetation. Thus, a good estimate of ET is required for sustainable management of water resources. However, planning at the river basin scale is challenging as it involves considering an area which may be several thousand square kilometres in size, and as such, there can be considerable variation in both climate and land use across the region. Measurements of ET are rarely available, and are unlikely to be sufficient to describe either the spatial variation or the influence of land use on the ET regime. In the absence of measurements using pan evaporimeters, an alternative solution is to use mathematical models or by using lysimeters or directly using Moderate Resolution Imaging Spectroradiometer Actual Evapotranspiration (MODIS AET) spatial and temporal data (Mu et al. 2007). Liu et al. (2010) used Surface-Energy-Balance ET algorithm to estimate AET at a higher spatial resolution using Landsat 5 satellite images. The lowest AET was observed for developed urban areas and highest for open water bodies. Guerschman et al. (2009) developed an algorithm for estimating monthly AET based on surface reflectance from MODIS-Terra and interpolated climate data. The model was able to reproduce the main spatial and temporal patterns in AET across Australia. It is able to estimate AET in areas with significant direct evaporation, including lakes and floodplains. Rao et al. (2014) estimated the annual, seasonal and mean monthly AET for different districts of Kerala using MODIS AET. This study was taken up with an objective to study the spatial and temporal variation in rainfall and AET in low, medium and high rainfall districts in Telengana region of Andhra Pradesh (AP) to act as a tool for its resource management.

Study Area

AP is one of the 28 states of India, situated on the country's southeastern coast. It is India's fourth largest state by area (275,045 km^2) and fifth largest by population. Two major rivers, the Godavari and Krishna, flow across the state. The state comprises three regions, namely Telangana, Coastal Andhra and Rayalaseema. The state receives heavy rainfall during southwest monsoon (July to September). The three districts selected namely, Adilabad, Warangal and Mahabubnagar selected in this study lies in the Telengana region. Adilabad lies in Godavari basin, Mahabubhagar lies in Krishna basin and Warangal lies partially in these two basins. Adilabad District lies between 18°40' and 19°56'N latitude and 77°46' and 79°59'E longitude with an elevation ranged from 91 to 666 m above mean sea level (MSL). Warangal District lies between 18°37' and 17°19' N latitude and 78°48' and 80°42'E longitude with an elevation ranged from 48 to 747 m above MSL. Mahabubnagar District lies between 15°49' and 17°15'N latitude and 77°14' and 79°15'E longitude with an elevation ranged from 156 to 883 m above MSL (*Source:* SRTM DEM). Major portion of Adilabad is characterised by clayey soil with moderate to severe erosion; Warangal is characterised by loamy and clayey soil with slight to severe erosion; Mahabubnagar is characterised by clayey and gravelly loam soil with slight to severe erosion.

Methodology

Spatial and Temporal Variation of Rainfall

The daily rainfall for 41 years (1971-2011), minimum temperature and maximum temperature data for 37 years (1971-2007) available at National Institute on Climate Resilient Agriculture (NICRA), Central Research Institute for Dryland Agriculture (CRIDA), Hyderabad was used for the data analysis in three selected districts, namely Adilabad, Warangal and Mahabubnagar respectively. The annual, mean annual and mean monthly rainfall and number of rainy days per year were estimated. The temporal variation (decadal and total) in mean annual rainfall and number of rainy days was worked out using CV (coefficient of variation). Extreme events such as maximum temperature, minimum temperature and rainfall were also worked out. From hydrologic point of view, the daily rainfall events contributing to runoff is very important. Hence, number of rainy days, per cent rainfall and per cent rainy days contributed to rainfalls <25 mm/day, 25-50 mm/day, 50-75 mm/day, 75-100 mm/day and more than 100 mm/day were estimated. Spatial variation of rainfall was plotted using World Clim data.

Estimation of Actual Evapotranspiration (AET) Using MODIS Data

MODIS monthly ET data (MOD16A2) in HDF (hierarchical data format) with sinusoidal projection, 1 km resolution, was downloaded from Numerical Terradynamic Simulation Group (NTSG) site for the period 2000 to 2012 (*ftp://ftp.ntsg.umt.edu/pub/MODIS/NTSG_Products/MOD16/*). After reprojecting to polyconic projection, raster images were multiplied with scale factor of 0.1 to get the original ET value using raster calculator in ARCMAP 10.0. Later annual AET was calculated using cell statistics tool in spatial analyst extension of ARCMAP.

In order to examine the AET for different types of LULC, 12 types of LULC were selected in the study area including agriculture lands (different types of crop land), forests (deciduous forest, scrub/degraded forest), grassland, wastelands and scrub lands. LULC map for the period 2004-2011 (1:50,000 scale) obtained from National Remote Sensing Centre (NRSC) Hyderabad, was used for estimating weighted average of AET. The disadvantage of using MODIS AET is that it gives filled values for build ups and water bodies. So MODIS data are suitable for classifying only cropped land, fallow land, wasteland, grassland and forest land. The AET values corresponding to water bodies and build up were masked since it uses filled values. The weighted average of AET for each zone was calculated year-wise using zonal statistics in Arc Map. The LULC for the period 2000-2003 was considered to be same as that of the year 2004 due to insufficient data availability. In order to understand the mean monthly variations of AET, the monthly MODIS AET corresponding to different LULC for 2000-2011 was determined and the mean monthly AET was estimated. The mean monthly and mean annual AET corresponding to three districts of AP was generated for the years 2000 to 2011 in order to determine the spatial variation of AET among three districts in comparison with the rainfall.

Spatial Variation of AET with Rainfall and Groundwater Status

The spatial variation AET during normal and dry years for all the three selected districts was compared. The spatial variation of mean annual AET in Warangal District with mean annual rainfall (*Source*: World Clim) groundwater status (*Source:* CGWB) was also compared.

Results and Discussion

Temporal and Spatial Variation of Rainfall

The rainfall data analysis indicated that out of the three selected districts, the highest mean annual rainfall for the last 41 years (1971-2011) occurred in Adilabad (1089 mm), followed by Warangal (990 mm) and Mahabubnagar (730 mm). The mean monthly rainfall during June, July and August was very high in Adilabad and Warangal compared to Mahabubnagar. The mean annual rainfall for the period 1971-1999 distributed in Adilabad, Warangal and Mahabubhagar were 1147, 994 and 725 mm respectively, whereas for the period 2000-2011 was 948, 979 and 743 mm respectively. During the last decade, the mean annual rainfall in Adilabad and

Warangal reduced, whereas Mahabubnagar increased. Also, the mean annual rainfall in Warangal is more than the rainfall in Adilabad. The temporal variation in mean annual rainfall was highest in Adilabad (CV=25.4%) followed by Mahabubhagar (CV=23.8%) and Warangal (CV=20.5%) (Table 6.1). Even though highest annual rainfall (1971-2011) occurred in Adilabad, more number of rainy days was observed in Warangal— showed the uniform distribution of rainfall in Warangal. The temporal variation in rainy days was highest in Mahabubhagar (CV=25%), followed by Warangal (CV=20.5%) and Adilabad (CV=17%). The decadal variation in rainy days almost doubled in the last decade in Warangal and Mahabubnagar Districts (Table 6.1). Spatial variation of rainfall was more in Mahabubnagar and Warangal Districts.

Extreme Rainfall and Temperature

Extreme daily rainfall events were more in Adilabad up to 1995. In 2006, both Warangal and Mahabubnagar experienced extreme rainfall. Extreme daily maximum temperature was highest in Adilabad and lowest in Mahabubnagar, whereas extreme daily minimum temperature was lowest in Adilabad and highest in Mahabubnagar. These results showed that Adilabad experiences more extreme weather events (1971-2007). The variation in extreme minimum temperature was more in all the three districts compared to the variation in extreme maximum temperature (Table 6.1).

Rainfall Thresholds

Rainfall distribution which contributed the runoff (1971-2012) is very important from hydrological point of view (Table 6.2). Rainfall less than 25 mm/day was highest in Warangal showing the uniform distribution of rainfall in Warangal. Rainfall more than 25 mm/day was more in Adilabad and Mahabubnagar compared to Warangal. In Adilabad, 59.3 per cent of the rainfall received during 1971-2011 was distributed from rainfall <25 mm/day, 22.7 per cent from 25 to <50 mm/day, 9.5 per cent from 50 to <75 mm/day, 4.3 per cent from 75 to <100 mm/day and 4.1 per cent from rainfall intensities >=100 mm/day. In Warangal, 73.9 per cent of the rainfall received was distributed from rainfall <25 mm/day, 18.9 per

Table 6.1

Decadal and Overall Variation (CV) in Annual Rainfall, Rainy Days and Extreme Maximum and Extreme Minimum Temperature in the Selected Districts

	Adilabad				*Warangal*				*Mahabubnagar*			
	Annual Rainfall	*Temperature*		*Rainy Days*	*Annual Rainfall*	*Temperature*		*Rainy Days*	*Annual Rainfall*	*Temperature*		*Rainy Days*
		Max	*Min*			*Max*	*Min*			*Max*	*Min*	
1971-1980	22.84	1.96	11.55	16.49	20.81	2.57	8.73	17.15	29.20	1.95	8.17	20.99
1981-1990	31.28	2.52	09.00	14.67	23.40	2.28	5.88	15.71	20.91	2.26	5.58	17.56
1991-2000	19.97	1.50	06.12	16.21	13.00	1.64	7.13	08.75	27.92	1.29	6.31	18.11
2000-2011	23.18	2.28	07.48	22.80	26.23	2.67	4.56	36.04	22.84	2.02	6.45	36.79
1971-2011	25.40	2.04	08.86	17.12	20.50	2.23	6.85	20.48	23.80	1.99	6.80	25.09

Source: World Clim.

Table 6.2

Daily Rainfall Thresholds in Adilabad, Warangal and Mahabubnagar (1971-2011)

Year	No. of Days Rf:<25 mm/day			No. of Days Rf: 25 <50 mm/day			No. of Days Rf: 50 <75 mm/day			No. of Days Rf: 75 <100 mm/day			No. of Days Rf: >= 100 mm/day		
	Adilabad	Warangal	Mahabubnagar	Adilabad	Warangal	Mahabubnagar	Adilabad	Warangal	Mahabubnagar	Adilabad	Warangal	Mahabubnagar	Adilabad	Warangal	Mahabubnagar
1	2	3	4	5	6	7	8	9	10	11	12	13	14	15	16
1971	73	80	37	4	3	4	0	0	0	0	0	0	0	0	0
1972	51	60	41	5	0	6	1	0	0	0	0	0	0	0	0
1973	62	80	48	13	4	9	3	0	0	0	0	0	0	0	0
1974	61	78	57	7	3	5	1	2	0	0	0	0	0	0	0
1975	91	100	70	5	8	6	1	0	2	1	1	0	0	0	0
1976	55	76	52	8	6	9	3	0	0	0	0	1	1	0	0
1977	62	84	54	8	1	6	2	1	0	1	1	0	0	0	0
1978	79	97	70	11	8	12	3	0	2	0	0	0	1	1	0
1979	59	66	54	5	1	2	3	1	1	1	2	0	0	0	0
1980	60	79	54	10	3	4	2	0	0	0	0	0	0	0	0
1981	66	78	62	12	4	7	2	1	1	0	1	0	0	0	0
1982	79	86	66	2	2	5	2	1	0	1	0	0	0	0	0
1983	83	85	62	11	7	9	2	0	2	0	0	2	1	0	0
1984	48	58	38	8	6	6	0	0	0	0	0	0	0	0	0
1985	69	77	44	6	1	3	1	2	0	0	0	0	0	0	0
1986	62	65	47	2	1	6	1	0	2	3	0	0	3	0	0
1987	71	85	69	3	4	2	1	0	1	0	1	1	0	0	0
1988	73	83	56	12	11	9	4	1	2	2	0	1	2	0	0

contd...

contd...

1	2	3	4	*5*	*6*	*7*	*8*	*9*	*10*	*11*	*12*	*13*	*14*	*15*	*16*
1989	63	69	55	6	6	8	2	0	1	3	1	1	1	0	1
1990	66	101	60	10	6	12	5	0	0	2	0	0	2	0	0
1991	50	80	51	7	5	4	2	1	2	0	1	0	0	0	0
1992	46	74	49	11	2	8	6	0	0	1	0	0	0	0	0
1993	55	82	47	12	1	3	2	0	0	0	1	0	0	0	0
1994	62	98	41	11	1	7	2	1	1	0	0	0	1	0	0
1995	61	80	70	13	5	11	4	1	0	3	0	0	1	0	0
1996	74	86	68	8	3	6	1	0	1	0	0	0	0	0	0
1997	82	86	59	6	2	4	1	0	0	0	0	0	0	0	0
1998	88	62	71	9	10	2	1	2	0	0	1	1	0	0	0
1999	86	91	49	7	1	3	0	0	0	1	1	0	0	0	0
2000	63	77	55	8	4	7	3	0	2	0	0	1	0	0	0
2001	61	35	45	11	8	6	0	0	1	1	0	3	0	0	0
2002	50	39	38	5	2	8	2	0	2	0	0	0	0	0	0
2003	69	49	43	9	5	6	1	0	2	0	0	0	0	0	0
2004	66	74	57	1	4	4	0	0	0	0	0	0	0	0	0
2005	118	129	124	0	0	0	0	0	0	0	0	0	0	0	0
2006	55	64	58	5	6	8	3	2	0	2	0	1	1	0	1
2007	57	82	70	6	7	6	0	0	1	0	0	0	0	0	0
2008	78	93	91	6	8	12	0	0	2	1	0	2	0	0	0
2009	62	56	47	5	6	2	0	2	1	0	0	0	0	0	0
2010	69	88	50	6	3	7	3	0	3	0	0	0	1	0	0
2011	59	73	46	7	1	8	2	0	1	0	0	0	0	0	0
Total	2744	3185	2325	301	169	252	72	18	33	23	11	14	15	1	2

cent from 25 to <50 mm/day, 3.5 per cent from 50 to <75 mm/day, 3.3 per cent from 75 to <100 mm/day and 0.4 per cent from rainfall intensities >=100 mm/day. In Mahabubnagar, 70.9 per cent of the rainfall received was distributed from rainfall <25 mm/day, 20.4 per cent from 25 to <50 mm/day, 5.0 per cent from 50 to <75 mm/day, 3.0 per cent from 75 to <100 mm/day and 0.7 per cent from rainfall intensities >=100 mm/day. Overall, 19-22 per cent of the rainfall quantity was received from 25 to 50 mm/day rainfall, 4-9.5 per cent rainfall from 50 to 75 mm/day rainfall, 3-4 per cent of rainfall from 75 to 100 mm/day and 0.4-4 per cent rainfall from more than 100 mm/day rainfall in all the three districts. The percentages 94.0, 88.5 and 87.0 per cent of the rainy days in Warangal, Mahabubnagar and Adilabad were from less intensity rainfall (<25 mm/day) (Table 6.3). The black soil along with the high rainfall events in Adilabad is contributing more surface runoff in Adilabad. This shows the scope for rainwater harvesting structures (ponds, check dams) for the critical irrigation in Adilabad.

Spatial and Temporal Variability of the Annual AET

Spatial variation of mean annual MODIS AET for the three districts during 2012 shows that in annual AET it was highest in Warangal and lowest in Mahabubnagar. The AET for the period 2000-2012 was extracted after overlaying the LULC map of three districts. The weighted average of AET for each LULC was calculated year-wise using zonal statistics in Arc Map. The results indicated that AET for forest was high, followed by grassland and wasteland, and it reached its peak in October and declined sharply. Lower AET was noted for crop land and scrub land and it reached its peak in July and declined slowly. This indicates that AET was controlled by the type of LULC and water availability at the same time. The spatial variation of AET during normal and dry years showed that the annual AET increased during dry year in Adilabad and Mahabubnagar, whereas it decreased in Warangal. This may be due to the insufficient availability of water in Warangal during the dry year. Significant variation in mean monthly AET was observed from low to high rainfall regions. The AET from low rainfall regions reached its minimum in February, and had started to rise in June

Table 6.3

Rainy Days and Rainfall Corresponding to Different Rainfall Thresholds in the Three Selected Districts (1971-2011)

% Rainy Days			% Rainfall			Rainfall Threshold
Adilabad	*Warangal*	*Mahabubnagar*	*Adilabad*	*Warangal*	*Mahabubnagar*	
86.97	94.12	88.54	59.3	73.9	70.9	<25 mm/day
9.54	4.99	9.60	22.7	18.9	20.4	25 to <50 mm/day
2.28	0.53	1.26	9.5	3.5	5.0	50 to <75 mm/day
0.73	0.33	0.53	4.3	3.3	3.0	75 to <100 mm/day
0.48	0.03	0.08	4.1	0.4	0.7	>=100 mm/day

due to increased moisture resulted from rain. The AET reached its peak in August-September and then declined. In case of other two districts with more rainfall, the AET started increasing in June, reached its peak in October and then declined.

Spatial Variation of AET and Groundwater Status

Among the three districts of AP, the mean annual AET was higher than the mean annual rainfall (2000-2012) in Adilabad and Warangal Districts, whereas it was lower than the mean annual rainfall in Mahabubhagar (Table 6.4 and Table 6.5). This shows the possibility of extraction of groundwater or other sources for irrigation to meet the additional AET demand in Adilabad and Warangal Districts. Hence, the groundwater status map was overlayed with the AET map and mean annual rainfall for Warangal District. It was found that overexploitation of groundwater was more in regions where the mean annual rainfall was less. Also, lower AET was noted in overexploited regions.

Conclusions

In this study, the spatial and temporal variation of rainfall and AET for the three districts of AP was determined. The highest mean annual rainfall for the past 41 years occurred in Adilabad followed by Warangal, whereas more number of rainy days was observed in Warangal showing the uniform distribution of rainfall in Warangal. The temporal variation in mean annual rainfall was highest in Adilabad followed by Mahabubhagar, whereas the temporal variation in rainy days was highest in Mahabubhagar followed by Warangal. The decadal variation in rainy days almost doubled in the last decade. Extreme rainfall, extreme maximum and extreme minimum temperature showed that Adilabad experienced more extreme weather events. Spatial variation of mean annual rainfall was more in Mahabubhagar and Warangal Districts. The AET exhibited remarkable spatial variability. Among the three districts of AP, the mean annual AET was higher than the mean annual rainfall (2000-2012) in Adilabad and Warangal Districts, whereas it was lower than the mean annual rainfall in Mahabubhagar. This shows the possibility of extraction of groundwater or other sources

Table 6.4

Mean Monthly Rainfall (mm) for Selected Districts of Andhra Pradesh (2000-2011)

Districts	January	February	March	April	May	June	July	August	September	October	November	December	Mean Annual Rainfall
Adilabad	8	7	16	16	18	166	236	284	154	31	9	2	948
Mahabubhagar	5	8	15	20	48	106	113	160	159	90	18	3	743
Warangal	8	11	17	20	31	144	213	280	173	52	22	7	979

Table 6.5

Mean Monthly AET (mm) for Selected Districts of Andhra Pradesh (2000-2011)

Districts	January	February	March	April	May	June	July	August	September	October	November	December	Mean Annual AET
Adilabad	77	80	79	80	84	103	130	128	134	139	104	82	1221
Mahabubhagar	35	37	49	46	48	60	80	79	71	71	62	49	689
Warangal	79	76	75	74	80	94	122	121	121	131	108	89	1169

for irrigation to meet the additional AET demand in Adilabad and Warangal Districts. The overexploitation of groundwater was more in regions where the mean annual rainfall was less and lower AET was noted in overexploited regions.

References

Aggarwal, P.K. (2007). "Climate Change: Implications for Indian Agriculture", *Hydrology Review* 22: 37-46.

Goswami, B.N., V. Venugopal, D. Sengupta, M.S. Madhusoodanam and P.K. Xavier (2006). "Increasing Trends of Extreme Rain Events over India in a Warming Environment", *Science* 314: 1442.

Guerschman, J.P., A.I.J.M. Van Dijk, G. Mattersdorf, J. Beringer, L.B. Hutley, R. Leuning, R.C. Pipunic and B.S. Sherman (2009). "Scaling of Potential Evapotranspiration with MODIS Data Reproduces Flux Observations and Catchment Water Balance Observations Across Australia", *Journal of Hydrology* 369: 107.-19

Guhathakurta, P. (2005). "Long-range Monsoon Rainfall Prediction of 2005 for the Districts and Sub-division Kerala with Artificial Neural Network", *Current Science* 90: 773.

Kumar, S. (2011). ''Land Use Land Cover Change and Atmospheric Feedback: Impact on Regional Water Resources.'', Ph.D thesis, Purdue University. Available online at *http://docs.lib.purdue.edu/dissertations/AAI3477689/*

Kumar, Vijay, Sharad K. Jain and Yatveer Singh (2010). "Analysis of Long-term Rainfall Trends in India", *Hydrological Sciences Journal* 55 (4): 484-96.

Liu, W., Y. Hong, S.I. Khan, M. Huang, B. Vieux, S. Caliskan and T. Grout (2010). "Actual Evapotranspiration Estimation for Different Land Use and Land Cover in Urban Regions Using Landsat 5 Data", *Journal of Applied Remote Sensing* 4.

Mu, Q., F.A. Heinsch, M. Zhao, S.W. Running (2007). "Development of a Global Evapotranspiration Algorithm based on MODIS and Global Meteorology Data", *Remote Sensing of Environment* 111: 519–36.

Rajeevan, M., J. Bhate and A.K. Jaswal (2008). "Analysis of Variability and Trends of Extreme Rainfall Events over India. Using 104 years of Gridded Daily Rainfall Data, Geophys", *Res. Lett*.

Rao, K.V., R. Rejani, G.S. Prathyusha Kranthi and J.V.N.S. Prasad (2014). "Estimation of Actual Evapotranspiration for Different Land Classes in Kerala Using MODIS Datas", *Compendium of the National Seminar on 'Spatial Technologies for Watershed Planning* during 8-9 January 2014. Thiruvananthapuram, Kerala: Kerala State Land Use Board. pp.54-64.

Teuling, A.J., M. Hirschi, A. Ohmura, M. Wild, M. Reichstein, P. Ciais, N. Buchmann, C. Ammann, L. Montagnani, A.D. Richardson, G. Wohlfahrt and S.I. Seneviratne (2009). "A Regional Perspective on Trends in Continental Evaporation," *Geophysical Research Letters* 36(2).

7

T. RAMANJALI, R. NARESHA,
J. NARENDER and D. ANIL

The Concept of Green Buildings Towards Sustainable Development

Introduction

A robust high performance buildings movement to rethink the built environment is rapidly emerging and affecting the design, construction and operation of new buildings; changing the renovation process for existing buildings; and reshaping cities and communities. The terminology used here to describe the new type of facilities resulting from this rethinking is high performance green buildings.

Green building designs and standards are developed to improve building operation energy and embodied energy efficiencies, and minimise energy and wastes (Kwok et al. 2011). Over the last two decades the construction industry has made efforts to develop green building practices (Gluch 2006). Many factors are promoting the rapid development of green buildings, including the increasingly serious environmental problems, the introduction and development of a variety of green building technologies, the successive implementation of accompanying 'green building evaluation criteria,' and other relevant policies and regulations.

Chatterjee (2009) defined the 'green building practice' as a process to create buildings and infrastructure in such a way that minimise the use of resources, reduce harmful effects on the ecology and create better environments for occupants. Green buildings exhibit a high level of environmental, economic and engineering performance. Green building concept has been adopted by many nations as the best way forward in preserving resources and sustaining environment (Al-Kaabi et al. 2009).

Green Building Initiatives in India

In India, the concept of green building came into being with the Confederation of Indian Industry (CII)-Godrej initiative in 2006, when the Indian Green Building Council was formed. The vision of the council is to serve as a nodal point to facilitate green building activities in India.

Green Building and Sustainable Development

Sustainable development is a major concern, and embodies both environmental protection and management. Generally, sustainable development concerns attitudes and judgement to help ensure long-term ecological, social and economic growth in society. The term 'sustainable construction' is used to describe the application of sustainable development in construction. To realise sustainable development in the construction industry, green building assessment system, this is designed to promote environmental awareness amongst built environment. There are a number of motives to building green, including environmental, economic and social benefits. Conservation of the ecosystems and biodiversity, improvement of the air and water quality, reduction of the amount of solid waste, natural resource conservation and depletion are considered to be the benefits for the environment.

Impact of Green Building on Climate Change

As buildings have different functions, climate change impact assessment studies must be tailored towards the specific needs and requirements at hand. The impact of climate change on buildings is deeply intertwined with consequences for the building occupants and key processes that take place in those buildings. It affects both the life of an individual and society as a whole as well as the environment and the natural ecological balance. It creates a logistical base in other economic sectors and describes economic development and culture of the country. Thus, reasonable usage of ecological systems and recoverable natural resources, that can reduce energy consumption and contribute to sustainable development of the country whilst ensuring quality of buildings and the environment, has become the essential task of the construction industry.

Green Buildings and Agriculture

In order to make the construction of green buildings cost-effective, the agricultural wastes, e.g., plant residues, should be used as sand materials. This concept has been also supported by Barreca (2012) who stated that the utilisation of local material in rural buildings minimises the energy cost for its transport as well as its environmental impact because when the building is demolished, the material is reintroduced into the environmental system. Therefore, the green building and agriculture are interdependent. Precisely, the agricultural wastes and the bio wastes (e.g., plant residues) can be used to make sustainable and recyclable green building materials (e.g., producing bio-cement, moulding plant residues) on the one hand and green buildings can provide sustainable agricultural structures on the other hand. Most of the green building materials should enter the natural cycle, i.e., originate from the nature and turn back into the nature where it will break down. In order to invest this interdependence, the role of agricultural and biological engineers, in the green building context, should be defined. Therefore, the role of agricultural and biological engineers—in cooperation with other disciplines can be elaborated as follows: (i) developing farm green building assessment and rating system; (ii) investigating the local agricultural materials that can be used as green building materials, e.g., giant reed, straw, clay etc.; (iii) manufacturing bio-materials, e.g., extracting bio-silica from plants, to be used for fabricating green building materials, e.g. bio-cement, eco-cement and green concrete; (iv) implementing the guidelines of green buildings when constructing new farm buildings; and (v) retrofitting old farm buildings to fulfil the green building criteria.

Integrated Technologies with Green Buildings

Zhai et al. (2007, 2008) designed and constructed a solar-powered integrated energy system including heating, air conditioning, natural ventilation and hot water supply. It is used for heating in winter, cooling in summer, natural ventilation in spring and autumn, hot water supply in all the year. The whole system is controlled by an industrial control computer and operates automatically. The experimental investigation validated the practical effective operation of the adsorption cooling-based air-conditioning system. They showed that for new buildings, solar collectors can be mounted on awnings besides roofs, on condition that solar systems become part of the general green building design. The solar-powered integrated energy system has the advantage of high utilisation ratio with different functions according to different seasons.

Ali and Al Nsairat (2009) developed a green building assessment tool (SABA Green Building Rating System) which is a computer-based programme that considers environmental, social and economical perspectives.

Challenges and Obstacles

The studies on agricultural buildings have focused on research points that form a small part of the whole green building concept. Such studies need to be integrated together to make the baseline and the first milestone on the way to apply the green building concept in agricultural buildings. Research projects should be developed on the implementation of green buildings in agriculture. The existing livestock barns and greenhouses do not comply with the green building concept as they miss some or most of the properties that formulate the green building aspect. Hence, the implementation of the green building concept in agricultural buildings is still limited, and therefore, should be conceptualised and initiated. Some studies implemented similar approaches to that adopted by green buildings, but they neither fully addressed the concept nor achieved its core.

Studies need to be integrated together to make the baseline and the first milestone on the way to apply green building concept in agricultural buildings. Research projects should be developed on the implementation of green buildings in agriculture.

Implemented Research Methodologies

A methodology is to consolidate the current foci of sustainable architecture. The research methods range from theoretical discussions of the usefulness of environmental management tools and questionnaire study on environmental management in the construction industry, to text analytical studies of media's representation of green building and field studies on environmental management in construction projects (Gluch 2006). Retzlaff (2008) conducted a content analysis of the system documentation for several building assessment systems and interviewed the administrators of the systems. Chau et al. (2010) mentioned that as the number and complexities of green building developments are mainly driven by market demands, understanding of end-user behaviours towards their development eventually should play a crucial role on determining their successes.

However, very few studies have been attempted to explore end-user behaviours towards green building development. They applied discrete choice experiments to reveal whether end-user with green experience will have different preference and willingness-to-pay values for enhancements on various aspects of environmental performance in green buildings. Generally, both green and conventional end-users have strong preferences and are willing to pay more for improving various aspects of environmental performance in green building developments. They are found to be willing to pay more for energy conservation, than indoor air quality improvement, noise level reduction, landscape area enlargement or water conservation. They found no significant differences in the preferences between green and conventional end-users for energy conservation, indoor air quality improvement, indoor noise reduction or water conservation. However, green end-users are willing to pay significantly less than the conventional end-users for enlarging the landscape area within a green building development, despite it was perceived by green end-users as one of the major elements that differentiate a green from a conventional development.

Conclusion

The green building and agriculture are interdependent. Precisely, the agricultural wastes and the bio-wastes can be used to make sustainable and recyclable green building materials on one hand and green buildings can provide sustainable agricultural structures on the other.

The studies on agricultural buildings have focused on research points that form a small part of the whole green building concept. Such studies need to be integrated together to make the baseline and the first milestone on the way to apply the green building concept in agricultural buildings. Research projects should be developed on the implementation of green buildings in agriculture.

Sustainable building involves solutions and practices that increase the efficiency of buildings, reducing energy, water and other natural resource consumption. Management processes help in developing green buildings, reduces power-intensity of buildings, their construction and negative effects on human health and the environment. It is achieved by selecting appropriate architectural and structural solutions, along with correct positioning of the building, so as to optimise the building in design, construction, maintenance and demolition using resources and assessing it through the whole life cycle—construction, maintenance and demolition stages.

References

Ali, H.H. and S.F. Al Nsairat (2009). "Developing a Green Building Assessment Tool for Developing Countries—Case of Jordan", *Building and Environment* 44(5): 1053-64.

Al-Kaabi, N.S., H.D. Imran, A.A. Al-Harmoudi, A.S. Al-Maamari, I.N. Al-Amirah and B.N. Rajab (2009). "An Application Model for Green Building Implementation: The Civil Engineer's Role." *Proceedings of the 12th International Conference on Civil, Structural and Environmental Engineering Computing*, 1-4 September. Funchal, Madeira, Portugal.

Barreca, F. (2012). "Use of Giant Reed Arundo Donax L. in Rural Constructions", *Agricultural Engineering International: CIGR Journal* 14(3): 46-52.

Chatterjee, A.K. (2009). "Sustainable Construction and Green Buildings on the Foundation of Building Ecology", *Indian Concrete Journal* 83(5): 27-30.

Chau, C.K., M.S. Tse and K.Y. Chung (2010). "A Choice Experiment to Estimate the Effect of Green Experience on Preferences and Willingness-to-pay for Green Building Attributes", *Building and Environment* 45(11): 2553-61.

Gluch, P. (2006). "Building Green Perspectives on Environmental Management in Construction." *PhD dissertation.* Gothenburg, Sweden: Chalmers University of Technology.

Kwok, K.Y.G., C. Statz and W.K.O. Chong (2011). "Carbon Emission Modelling for Green Building: A Comprehensive Study of Methodologies." Proceedings of the International Conference *on Sustainable Design and Construction (ICSDC): Integrating Sustainability Practices in the Construction Industry*. pp.9-17. 23-25 March. Kansas and USA.

Retzlaff, R.C. (2008). "Green Building Assessment Systems: A Framework and Comparison for Planners", *Journal of the American Planning Association* 74(4): 505-19.

Zhai, X.Q., R.Z. Wang, Y.J. Dai, J. Wu, Y. Xu and Q. Ma (2007). "Solar Integrated Energy System for a Green Building", *Energy and Buildings* 39(8): 985–93.

Zhai, X.Q., R.Z. Wang, Y.J. Dai, J.Y. Wu and Q. Ma (2008). "Experience on Integration of Solar Thermal Technologies with Green Buildings", *Renewable Energy* 33(8): 1904–10.

Section 3
Sustainable Agriculture

8

R. NARESHA, T. RAMANJALI,
J. NARENDER and G. PADMAJA

Impact of Climate Change on Agriculture

Introduction

Agriculture is an economic activity that is highly dependent upon weather and climate in order to produce food and fibre necessary to sustain human life. Not surprisingly, agriculture is deemed to be an economic activity that is expected to be vulnerable to climate variability and change.

Agriculture, rural livelihoods, sustainable management of natural resources and food security are inextricably linked within the development and climate change challenges. Successful adaptation and mitigation in agriculture can only be achieved within the ecological, economic and social sustainability goals set forth by the World Food Summit, the Millennium Development Goals (MDGs) and the United Nations Framework Convention on Climate Change (UNFCCC).

The international community faces great challenges in the coming decades including reining in global climate change, ensuring food security for the growing population, and promoting sustainable development. Changes in the agriculture sector are essential to meeting these challenges. Agriculture provides the main source of livelihood for the poor in developing countries, and improving agricultural productivity is critical to achieving food security as well as most of the targets specified under the MDGs (Rosegrant *et al.*, 2006). Agriculture also contributes a significant share (14%) of GHG (greenhouse gas) emissions, more if related land-use change (particularly deforestation) is included (World Resources

Institute [WRI] 2010). At the same time, long-term changes in average temperatures, precipitation and climate variability threaten agricultural production, food security and livelihoods of the poor. While mitigation of GHG emissions can lessen the impact of climate change, adaptation to climate change will be essential to ensure food security and protect the livelihoods of poor farmers.

Climate Change and Agriculture

There is a great deal of uncertainty about climate change, but there are some certainties. Average global temperatures are rising, and will continue to rise over the coming decades, whatever mitigation measures are taken, because of stocks of GHGs already in the atmosphere. These rising temperatures are already having measurable impacts, on glaciers and ice caps, sea levels, and rainfall patterns, and these impacts will also increase over the next decades. The ultimate temperature rise will depend on mitigation measures put in place to limit emissions over the coming years.

Soils can be better managed to store carbon, while agro forestry is also an underutilised mitigation option. Biofuels are a complex issue at the moment. They have potential for reducing greenhouse gas emissions by replacing fossil fuels, but their production has its own environmental costs and may compete with that of food and feed. Research into biofuel development should clarify the issues, and allow production of biofuels in an environmentally sustainable way that also benefits the poor. These, and other promising mitigation options linked to agriculture, land use and natural resources management, are explored further below.

Many pests and diseases of crops, animals and humans are sensitive to climate, and we can expect these to change in currently unpredictable ways. Some will become prevalent in areas where they were previously unknown, when the climate becomes favourable in those areas. The danger is that there is usually low immunity to a disease, and poor knowledge of pest or disease management, in areas where they have not occurred before.

Natural ecosystems are equally at risk from climate change. The natural environment is not static, and ecosystems have evolved and adapted to gradually changing environmental conditions throughout history. A fundamental challenge of climate change is its rapid pace, and plants and animals that cannot quickly adapt to new conditions or relocate to new areas will become extinct. Along with destabilising vital ecosystems, this also erodes the genetic base for future crops and livestock.

Fishing communities are equally vulnerable. Water temperatures and acidity will change, as well as sea levels, upwelling and ocean currents. Impacts on marine ecosystems, particularly coral reefs, could be devastating. While aquaculture will be challenged by increasing temperatures and extreme events, marine aquaculture will be challenged by environmental hazards such as harmful algal blooms.

Agricultural practices that have benefits for climate change adaptation or productivity enhancement may increase GHG emissions. For instance, expanding agricultural production, a reported adaptation strategy, can increase total farm production and provide benefits in terms of adaptation; however, the cultivation of new lands that were previously under forest, grasslands or other non-agricultural vegetation can release additional GHGs. In many cases, fertiliser application can also result in increased emissions. This is the case in some regions, such as Asia, where fertiliser application rates are already high. However, fertiliser use in much of Sub-Saharan Africa (SSA) is so low that increased application in these areas is likely to mitigate climate change rather than increase emissions. In fact, the benefits of appropriate fertiliser use in SSA are immense—a study of maize and bean yields over an 18-year period in Kenya showed dramatic increases when crop residues were retained and fertiliser and manure were applied to the soils. Therefore, increased fertiliser application (in conjunction with soil fertility management) in this context reduces soil mining and supports mitigation, adaptation and agricultural productivity.

Whatever mitigation efforts are made, climate change is already happening and temperatures will continue to rise during the coming years. It is not known exactly what will happen where, but the impacts we are beginning to see will intensify. These changes will happen at the same time as, and will interact with, the impacts of other global trends such as population growth, urbanisation, increasing demands for water, over-exploitation of ecosystems, and shifts in world economics. If we are to understand real-life impacts, and develop meaningful responses, we must look at climate change and agriculture within this global system. The challenge is dynamic and multifactorial; the responses—the adaptation options that will allow people to manage this challenge—will have to match.

Mechanism for Climatic Impacts on Crops

Hulme (1996) describes four ways in which climate would have a physical effect on crops. First, changes in temperature and precipitation will alter the distribution of agro-ecological zones. Changes in soil moisture and content, and the timing and length of growing seasons will be affected in various ways in different parts of the world. In contrast, in lower latitudes, it is expected that higher temperatures will adversely affect growing conditions, especially in areas where temperatures are close to or at the optimal level for crop growth to begin with. Irrigation availability and demand will also be affected by both changes in temperature and precipitation.

Reduction in precipitation is likely to intensify further aquifer exploitation for agriculture and place additional burdens on other surface and groundwater resources from non-agricultural use (such as industrial and municipal needs). An increase of potential evapotranspiration is likely to intensify drought stress, especially in the semi-arid tropics and sub-tropics (Hillel and Rosenzweig 1989).

Agriculture and Sustainability

Most studies take the effect of climate change into account while all other factors are held constant. While past experience has indicated that the agricultural sector is able to respond and adapt quickly to changes in climate, government will play a role in how the sector adapts and how quickly.

Government compensation, subsidy and assistance are programmes which the government uses to offer rural communities some form of economic and social security. However, if farmers on an individual level are always able to access government assistance when faced with climatic catastrophe, they do not have an incentive to change their practices so that they match the climatic reality since the government will always bail them out. A balance needs to be found between government programmes that help in emergencies and act as short-term coping strategies *versus* government programmes that encourage their use as an adaptive strategy.

The changing climate on the Canadian prairies could have a significant role to play on the population. Assuming that agriculture moves to more diversity, there will be an increase in available jobs. This is based on the fact that cereal production on the prairies is much mechanised, while other production, such as livestock and speciality crops, is more dependent on labour. The counter argument is that if climate change has reduced the ability of the prairies to support humans, based on the quality and quantity of resources such as water, soil, flora and fauna, then out-migration can be expected.

Issues for Future Impacts and Adaptation

Adverse effects from climate change can be reduced through successful adaptation, which would likely be less than the cost of the impacts that would otherwise occur without adaptations. Further, we can anticipate that there could be significant implications arising from climate change for agriculture, both directly upon our own food production, and indirectly from the impacts upon competing agricultural regions in other countries. One prediction is that there could be benefits associated with climate change, with a potential increase in agricultural production as favourable agro-climatic conditions expand northward in the prairies. Such an expansion however would be tempered by the capacity of northern soils to sustain commercial agriculture.

Conclusion

Climate change promises serious negative impacts on agricultural systems. These same systems and the natural resources that support them are already under severe strain from over exploitation, the current climate and multiple other stresses. Many of the world's most vulnerable people depend directly on these systems for their food and livelihoods, and many countries, economies are also highly dependent on them. Agriculture is also adding to the climate change problem.

Global warming is now causing unprecedentedly rapid changes in the climate conditions that affect agriculture—much faster than crops can evolve on their own, and probably too fast for the traditional processes of trial-and-error adaptation by farmers. At the same time, the world's population will continue to grow through mid-century or later, increasing the demand for food just as climate change begins to depress yields. To adapt to the inescapable early states of climate change, it is essential to apply the rapidly developing resources of plant genetics and biotechnology to the creation of heat resistant, and perhaps drought-resistant, crops and cultivars.

We are at a crossroads in the development of our planet. The decisions we make now, for agriculture and natural resources as well as for other sectors, may prove to be the most important decisions humankind ever collectively makes. The two groups that stand to lose the most if the wrong decisions are made—the world's poor and future generations—are, ironically, the same groups who are least responsible for climate change and who have the least voice in the debate.

Food security is one of the basic human rights that are in jeopardy. Today, an estimated one billion people—more than one person in six—do not enjoy that right; without significant changes to agriculture and other natural resource-based systems, hunger and poverty will be perpetuated long into the future and affect many more.

Even without climate change, we have a moral imperative to turn this knowledge into action. Climate change adds urgency to the situation, but it also provides an opportunity. The products of agricultural research are ready to be implemented in adaptation and mitigation strategies that will help people build successful livelihoods despite changing conditions.

References

Consultative Group on International Agricultural Research (CGIAR) (2009). *Climate, Agriculture and Food Security: A Strategy for Change. www.cgiar.org.*

Food and Agriculture Organisation of the United Nations (FAO) (2012). *Climate Change Adaptation and Mitigation Challenges and Opportunities in the Food Sector.*

Hillel, D., and C. Rosenzweig (1989). "The Greenhouse Effect and its Implications Regarding Global Agriculture", *Massachusetts Agr. Exp. Stn. Res. Bull.* 724. Amherst.

Hulme, M. (1996). "Global Warming", *Progress in Physical Geography* 20(2): 216-23 (R).

Rosegrant, M.W., C. Ringler, T. Benson, X. Diao, D. Resnick, J. Thurlow, M. Torero *et al.* (2006). "Agriculture and Achieving the Millennium Development Goals." *World Bank Report* No. 32729-GLB. Washington, DC: World Bank.

World Resources Institute (WRI) (2010). *Climate Analysis Indicators Tool*, Version 5.0. Accessed on April 7. *http://cait.wri.org/.*

9

B.N. SANDEEP NAYAK, M.D. MUJEEB KHAN,
K. MOSHA and P. PRASUNA RANI

Plant Density and Weed Management Effect on the Productivity of Drum Seeded Rice (*Oryza sativa L.*)

Introduction

Rice (*Oryza sativa L.*) is a staple food for nearly half of the world's seven billion population. However, more than 90 per cent of rice is consumed in Asia, where it is a staple food for a majority of the population, including the 560 million hungry people in the region (Mohanty 2013). Globally, India stands first in area for rice production and second in production after China. It is also a staple food for more than 65 per cent of the Indian population and accounts for more than 42 per cent of food production.

The area under direct-seeded rice is increasing as farmers in India seek higher productivity and profitability to overcome increasing costs and scarcity of farm labour. One of the major reasons for non-remunerative rice production in recent times is augmented cost of cultivation because of scarce and costly farm labour during the peak period of farm operations. Establishing rice by transplanting is labour intensive and increasingly difficult due to higher cost and shortage of labour. Inadequate plant density with hired labour for transplanting is the major lacuna in this method (Ram et al. 2006).

Drum seeding is an alternative method to transplanting. It reduces labour requirement and performs as good as transplanting method at many places (Yadav and Singh 2006). However, drum seeding method is subjected to severe weed infestation than conventionally puddled transplanted rice because of the absence of the size disparity between the crop and weed plants and the suppressive effect of standing water on weed growth at crop establishment.

Conventional methods of weed control such as hand weeding may not be possible in such areas where large labour force is needed at a given time. Dumping of high dosage of herbicides are not only affecting the soil flora and fauna, but also reducing the production capacity of soil as well as plants by various ways. Low dosage and high efficacy herbicides not only are environment friendly chemicals but also economical in managing weeds.

Material and Methods

A field experiment entitled "Plant Density and Weed Management Effect on the Productivity of Drum Seeded Rice (*Oryza sativa L.*)" was conducted at the Agricultural College Farm, Bapatla, on sandy loam soil during *kharif* 2012. The treatments consisted of combination of five drum seeder spacings (20×7cm, 20×10.5 cm, 20×14 cm, 20×17.5 cm, 20×24.5 cm, and manual planting (20×15 cm), with a rice plant density of 71, 47, 35, 28, 20 and 33 hills m^{-2}, respectively, and five weed management practices *viz.*, weedy check (W1), hand weeding at 20 and 40 DAS (W2), cono weeding twice at 20 and 40 DAS with modified cono weeder (W3), pre-emergence application of anilofos @ 0.375 kg a.i. ha^{-1} and post-emergence application of 2, 4 D salt @1.0 kg a.i. ha^{-1} at 25 DAS (W4), pre-emergence application of pendimethalin @1.0 kg a.i.. ha^{-1} post-emergence application of bispyribac sodium @ 20 g a.i. ha^{-1} 30 DAS (W5).

The trail was laid out in strip plot design and replicated thrice. The rice variety used was NLR-33358 (SOMASILA). Fertiliser was applied at the rate of 120:60:60 N: P2O5:K2O kg ha-1. Nitrogen was applied in two split doses at time of tillering and PI (panicle initiation) stage along with basal dose. Phosphorus and potassium was applied as basal.

Result and Discussion

Predominant weed flora in the experimental field were *Echinochloa colona, Echinochloa crusgalli* and *Cynodon dactylon* among grasses; *Cyperus rotundus, Cyperus difformis,* and *Fimbristylis miliaceae* among sedges; and *Eclipta alba, Ludwigia parviflora, Commelina benghalensis* among broad leaf weed in puddled condition.

Effect of Plant Density and Weed Management Practices on Weeds

Weed density and weed dry matter production were significantly lower with plant density of 71 hills m^{-2} as compared to other rice plant densities. Among weed management practices, the lowest weed density and weed dry matter production were recorded with cono weeding twice at 20 and 40 DAS (W3), which was on a par with hand weeding twice at 20 and 40 DAS (W2) at all stages of crop growth except at 20 DAS. However the weedy check treatment registered the highest weed density and dry matter production over all other treatments. Similar results were observed by several researchers, including Bhowmick et al. (2000) and Walia et al. (2008).

A significant interaction between rice plant density and weed management practices showed that all the treatments were significantly superior over weedy check in reducing weed dry matter production at all the crop growth stages. However, W2, W3 or W5 with higher rice plant densities performed excellently compared to other treatment combination.

The highest WCE (weed control efficiency) was recorded with plant density of 71 hills m^{-2} which was significantly superior compared to all other treatments. Among the weed management practices, weed control efficiency was significantly higher with weeding twice at 20 and 40 DAS with modified cono weeder. Among herbicidal treatments pre-emergence application of pendimethalin @ 1 kg a.i. ha^{-1}, followed by post-emergence application of bispyribac sodium @ 20 g kg ha^{-1}, was recorded significantly superior over application of pre-emergence application of anilofos @0.375 kg a.i. ha^{-1}, followed by post-emergence application of 2, 4 D Salt @1 kg a.i. ha^{-1}.

Effect of Plant Density and Weed Management Practices on Dry Matter Production of Rice

In general, there is linear increase in rice dry matter production up to flowering stage; there after it was stagnant. Among different plant densities, rice dry matter production was higher in thicker plant densities as compared to thinner densities. This increase was very prominent with higher dry matter production recorded in D1 (71 hills m^{-2}) at maximum tillering (156.4 g m^{-2}), PI (871.9 g m^{-2}) flowering (1316 g m^{-2}), which was significantly superior to other plant densities. Thinner plant density treatment with mere 20 hills m^{-2} (D5) recorded lowest rice dry matter production at maximum tillering (110.8 g m^{-2}), PI (481.7 g m^{-2}), and flowering (932.8 g m^{-2}) stages. The possible reason for this trend may be that increasing plant density increases LAI (leaf area index) of crops, which favours higher PAR (photosynthetic active radiation) interception resulting in higher dry matter production by crop. Lower weed infestation with higher plant densities further maintained the beneficial effects of increased rice dry matter production. These results confirmed the findings of Mahajan et al. (2010) and Zhao et al. (2010).

At PI, flowering and harvest stages manual (W2) or mechanical (W3) weeding methods performed better in increasing rice dry matter production. This might be owing to the fact that disturbing the soil by manual or mechanical means created well-aerated atmosphere for better growth of rice root system, which paved the way for better shoot growth due to better nutrient uptake at the later stages of crop growth. These results are akin to the findings of Yadav and Singh (2006).

A significant interaction at most of the rice crop growth stages showed that initially high rice plant density with pre-emergence and post-emergence herbicide (W5) combination resulted in higher rice dry matter production. During later stages of crop growth, high plant density with mechanical and manual weeding resulted in higher rice dry matter production. At all the stages, irrespective of variation in rice plant density, the weedy check treatment showed lowest rice dry matter production. The probable reason for this might be severe crop-weed competition that resulted in weak and lanky rice plant growth, thereby causing lowest rice dry matter production as compared to intervention of weed control by any of the weed management practices.

Effect of Plant Density and Weed Management Practices on Yield Attributes of Rice

Productive tillers per unit area were significantly increased with an increase in rice plant density with maximum number observed in 47 hills m^{-2} (D2) followed by D1. Maximum productive tillers (279 m^{-2}) recorded with a plant density of 47 hills m^{-2} (D2) were significantly higher compared to a plant density of 71 hills m^{-2} (D1), 35 hills m^{-2} (D3), 28 hills m^{-2} (D4), 20 hills m^{-2} (D5) and transplant (D6) 233, 171, 149, 92 and 196 m^{-2}, respectively. Lesser number of productive tillers in D1 than D2 may be due to inter- or intra-plant competition due to higher plant density leading to more mortality of normal tillers. However, lesser number of productive tillers in D3 to D6 treatments compared to D2 might be due to lesser number of normal tillers owing to reduced level of plant density.

Among weed management practices, cono weeding twice at 20 and 40 DAS (W3) recorded highest number of productive tillers (244 m^{-2}) and it was on a par with hand weeding twice at 20 and 40 DAS (W2). Lowest number of productive tillers (119 m^{-2}) was recorded with weedy check (W1) compared to other weed management practices.

Panicle length, test weight and total number of grains per panicle were significantly higher at a plant density of 20 hills m^{-2} over the rest. With regard to weed management treatments, cono weeding twice at 20 and 40 DAS recorded more productive tillers, panicle length, test weight and total number of grains panicle-1. In chemical control, pre-emergence application of pendimethalin and post-emergence application of bispyribac sodium (W5) showed superiority in yield attributes. Earlier Walia *et al.* (2008) and Yadav *et al.* (2009) also observed the effectiveness of bispyribac-sodium in controlling weeds without any phytotoxic effect on rice.

Effect of Plant Density and Weed Management Practices on Paddy Yield

Among various rice plant densities, a medium level population of 47 hills m^{-2} (D2) significantly increased the paddy yield over all other treatments except D1 treatments with a plant density of 71 hills m^{-2}. The highest paddy yield of 3,476 kg ha^{-1} was observed with a plant density of 47 hills m^{-2} and it was significantly superior to 35, 28, 20 drum seeded and 33 hills m^{-2} transplanted paddies. It was on a par with a paddy yield of 3,154 kg ha^{-1} with a plant density of 71 hills m^{-2} (D1). The manual transplant (D6) gave paddy of 3,085 kg ha^{-1} which was at par with the plant density of 71 and 35 hills m^{-2} drum seeded rice with 3,154 and 3,060 kg ha^{-1}, respectively. Mahajan et al. (2010) also demonstrated that rice grain yield increased with an increase in plant density to a certain level, further increase of plant density beyond optimum level had a negative effect owing acute inter- and intra-plant competition for available resources.

A lower plant density of 28 and 20 hills m^{-2} in D4 and D5 treatments reduced the paddy yield due to inadequate spacial occupation for optimal utilisation of *in situ* and externally applied resources there by negating the phenomenon law of constant described by Bond et al. (2005).

Among the weed management practices, significantly higher paddy yield (3,747 kg ha^{-1}) as compared to all other weed management practices was recorded by twice cono weeding at 20 and 40 DAS (W3), which was on a par with twice manual weeding 20 and 40 DAS (W2) treatments (3,570 kg ha^{-1}). The significant paddy yield (1,188 kg ha^{-1}) was recorded by the weedy check (W1) treatment. Among chemical methods of weed management, pre-emergence application of pendimethalin @1 kg a.i. ha^{-1}, followed by post-emergence application of bispyribac sodium @ 20 g a.i. ha^{-1} at 25 DAS (W5) was found better in increasing the yield over the pre-emergence application of anilofos @ 0.375 kg a.i. ha^{-1} followed by post-emergence application of 2,4.D sodium salt @ 1 kg a.i. ha^{-1} (W4). The increase in paddy grain yield cono weeding twice (W2), hand weeding twice (W3), application of pendimethalin followed by bispyribac-sodium (W5) and application of anilofos followed by 2,4 D sodium salt (W4) over weedy check (W1) was 67.4, 66.4, 63 and 56 per cent, respectively. Sequential application of pre-emergence followed by post-emergence herbicide proved better for prolonged period of controlling weeds to realise higher yields in rice. These results are in conformity with the finding of Bhowmick et al. (2000).

A significant interaction between rice plant densities and weed management practices showed that a treatment combination of D2×W3 gave the highest paddy yield of 4275 kg ha^{-1} which was significantly superior to all other treatment combination. Next best treatment combination is D2×W2 and D1×W3 with a grain yield of 4,142 kg ha^{-1} and 4,124 kg ha^{-1} and superior to all other treatment combination even when compared with transplanted paddy. These results clearly showed that medium to slightly higher plant densities above 33 hills m^{-2} with a combination of weed management technique, which will serve the dual purpose of controlling the first and second generation of both grassy and broad leaved weeds with an added advantage of soil pulverisation so as to enhance intermittent aeration, would be the better option to extract higher rice productivity through direct seeded method of drum seeding which is even better than the traditional system of transplanted paddies particularly under the situation of depleting manual labour scenario.

References

Bond, J.A., T.W. Walker, P.K. Bollich, C.H. Koger and P. Gerard (2005). "Seeding Rates for Stale Seedbed Rice Production in the Midsouthern United States", *Agronomy Journal* 97: 1560–63.

Bhowmick, M.K., R.K. Ghosh and D. Pal (2000). "Bio-efficacy of New Promising Herbicides Weed Management in Summer Rice", *Indian Journal of Weed Science* 32 (1&2): 35-38.

Mohanty, S. (2013). "Trends in Global Rice Consumption", *Rice Today:* 44-45.

Mahajan, G., M.S. Gill and K. Singh (2010). "Optimising Seed Rate to Suppress Weeds and to Increase Yield in Aerobic Direct-seeded Rice in Northwestern Indo-Gangetic Plains", *Journal of New Seeds* 11:125-38.

Ram, M., O.M. Hari, S.D. Dhiman and D.P. Nandal (2006). "Productivity and Economics of Rice (*Oryza sativa*)-Wheat (*Triticum aestivam*) Cropping System as Affected by Establishment Methods and Tillage Practices", *Indian Journal of Agronomy* 51(2): 77-80.

Walia, U.S., O. Singh, S. Nayyar and V. Sindhu (2008). "Performance of Post-emergence Application of Bispyribac in Dry-seeded Rice", *Indian Journal of Weed Science* 40(3&4): 157-60.

Yadav, V. and B. Singh (2006). "Effect of Crop Establishment Method and Weed-management Practice on Rice (*Oryza sativa* L.) and Associated Weeds", *Indian Journal of Agronomy* 51(4): 301-03.

Yadav, D.B., A. Yadav and S.S. Punia (2009). "Evaluation of Bispyribac-sodium for Weed Control in Transplanted Rice", *Indian Journal of Weed Science* 41 (1-2): 23-27.

Zhao, L., L. Wu, Y. Li, A. Sarkar, D. Zhu and N. Uphoff (2010). "Comparisons of Yield, Water Use Efficiency, and Soil Micro-bial Biomass as Affected by the System of Rice Intensification", *Communications in Soil Science and Plant Analysis* 41: 1-12.

10

A. RAKESH, V. HEMALATHA,
K.B. ESWARI and V. SWARNALATHA

Sustainable Agriculture

A Review

Introduction and Need of Sustainable Agriculture

Agriculture is an essential occupation needed to feed the enormously growing world's population. Due to new technologies, mechanisation, high amount of chemical use with the aim of increasing the production, the agriculture has changed compared to earlier. Although the modern agriculture is having benefits like: abundant food supply in the developed world, fresh fruits and vegetables available year-round, cheap food, luxury foods easily available around the world, effective food preservation technologies (refrigeration, freezing, canning, packaging), mechanisation produces high labour efficiency and availability of agricultural inputs for quick solutions to production problems, it is having negative effects like: continuing soil loss, food safety concerns, water pollution, air pollution, habitat loss, water depletion, economic instability, many diseases linked to agricultural chemicals, dependence on fossil fuels and global warming. These changes also influence the lives of labourers, small farmers and cause disruption of economic and social conditions in rural communities. Nowadays, everything has become inorganic even the land and water due to constant use of fertilisers, pesticides and other agriculture chemicals. As people want fast results and fast yields, they turn to inorganic ways. But by these actions in modern agriculture, future generations will lose the ability of exploring and utilising natural resources in farming.

'Sustainable agriculture' is an integrated system of plant and animal production practices that will satisfy human food and fibre needs, enhance environmental quality, make the most efficient use of non-renewable resources, sustain economic viability, and enhance quality of life. Sustainable crop production is a way of growing or raising food in an ecologically and ethically responsible manner. This includes agricultural and food production practices that do not harm the environment, that provide fair wages to workers and that support and sustain local communities. Sustainable crop production practices can lead to higher yields for long time with less inputs, which are environmentally hazardous.

Main Goals of Sustainable Agriculture

Sustainable agriculture has environmental, social and economic dimensions or goals—and all three must be considered together. Focusing on one or two in isolation will not give the desired results. Protecting and improving the natural environment are fundamental, and issues like climate change, energy, water scarcity, biodiversity and soil degradation need to be addressed. The social dimension covers labour rights and the health of communities, including access to and affordability of food, labour rights and community health. Food quality, safety and animal welfare are also important social aspects. On the economic side, sustainable agriculture is productive, efficient and competitive. The benefits should be seen in farm profitability, in thriving local economies and throughout the whole value chain.

Economic Sustainability

To be truly sustainable, a farm must be economically stable. Sustainable agriculture practices can have a positive economic impact on a farm. For example, diversifying the farm with several crops and markets helps to reduce financial risk. Over time, improved soil and water quality, as well as other environmental benefits from sustainable practices, may raise the value of the farm. Selling products directly in the local markets results in decreased transportation costs. Production costs can be variously affected by sustainable methods. The agriculture chemicals like pesticides and fertilisers costs are generally reduced on a sustainably managed farm because legumes and crop rotations tend to be less expensive than the synthetic chemicals. A farm in which the family savings or net worth is continuously increasing, the family debt is continuously decreasing and purchase of inputs is decreasing is said to be economically sustainable.

Environmental Sustainability

Environmental concerns are central to sustainable agriculture. Sustainable agriculture involves ecological friendly practices that have little or no adverse effect on natural ecosystems. Sustainable agriculture also has a positive effect on natural resources and wildlife. This can often mean taking measures to decreasing the damage (e.g., soil erosion or draining of wetlands) that have already occurred through harmful modern agricultural practices. Renewable natural resources are protected, recycled and even replaced in sustainable systems. Achieving a balanced ecosystem takes time. Making the transition to sustainable farming is a process that generally requires moving forward in step-wise manner. A farm in which there is no cultivable land left as bare land, water quality is good and rich in biological diversity is said to be environmentally sustainable.

Social Sustainability

Social sustainability refers to the quality of life for the men who work and live on the farm, as well as those in the local community. Fair treatment of workers, positive farm family relationships, personal interactions with consumers and choosing to purchase supplies locally rather than from distant market are some of the aspects focused in social sustainability. The farm supports the community and the community supports the farm in essential situations. A farm in which it gives support to other businesses and families in community, currency circulate within the local economy and farming continues to young people from parents is said to be socially sustainable.

Natural Resources Involved in Farming

Water, air and soil are the important natural resources in farming/agriculture. If we degrade the natural resources, the ability of future generations to produce will decrease. Sustainability is an attempt to provide the best outcomes for the human and natural environments both for present and future.

Water

Water is vital resource that helps agriculture and society to develop. It is a major limiting factor when mismanaged. Water is essential for life and is one of the important natural resources. A substantial increase in consumption by both the agricultural and industrial sectors raises concern on the future availability of water. Water scarcity has grave implications for agriculture, as almost 70 per cent of the world's surface water supplies are used for farming. Therefore, it is essential to protect this precious resource by implementing efficient water management techniques and sustainable practices like efficient irrigation systems, water harvesting and waste control.

Water Quality

The most important issues related to water quality involve salinisation and contamination of ground and surface waters by agriculture chemicals like pesticides, nitrates, etc. Soil salinity is the salt content in the soil and the process of increasing the salt content is known as salination. Naturally salt is present both in soil and water. Natural processes such as weathering of minerals can cause salination. Artificial processes such as irrigation are also responsible for salination. The excess accumulation of salts, typically higher at the soil surface, can result in salt-affected soils. From the water table rich in salts, salts may rise to surface of the soil and this results in salt accumulation due to evaporation. As soil salinity increases, salt effects can result in degradation of soils and vegetation. Salinisation as a process can result from: high levels of salt in water, movement of water table, climatic factors that favour accumulation, and human activities such as land clearing, excess use of agriculture chemicals like pesticides, fertilisers, etc., result in degradation of water quality.

Air

The quality of air is most affected by many agricultural practices. These include smoke from burning of agricultural waste, dust from fields during various cultural operations, drift from spraying of pesticides and nitrous oxide emissions from the use of nitrogen fertiliser. The measures to improve air quality include: incorporating crop residue into the soil, using appropriate levels of tillage, planting wind breaks and cover crops to reduce dust.

Soil

Soil erosion is a serious threat to our ability to produce adequate food. Management practices developed to keep soil erosion in control include: minimum or no tillage, managing irrigation to reduce run-off and keeping the soil covered with plants or mulch.

How to Maintain Sustainability in Agriculture

By the improvement in technology in agriculture during earlier decades, the production from agriculture has increased. This also results in damage to the environment, soil depletion, air pollution, etc. By following different practices, we can maintain sustainability of natural resources. These practices includes diversification of crop and livestock production, selection of crops well suited to the area of production, limited use of artificial/synthetic chemicals, etc.

Some principles involved in sustainable crop production include:

Multiple Cropping

Multiple cropping refers to growing more than one crop on the same field during the season. This technique makes effective use of inputs such as soil, water, fertiliser, etc. Thus output per unit area increases with manifold returns to the growers. Multiple cropping can be done in annual food crops, fodders, vegetables, fruit plants and perennial crops. With multiple cropping, incidence of crop failure owing to biotic agents is minimised. One crop may provide cover to the other against such agents through biological control. If legume crops are included in multiple cropping, they are very much helpful in increasing soil nutrient status by a process of nitrogen fixation. An important aspect of multiple cropping is the utilisation of nutrients more efficiently as the crops growing on the same piece of land simultaneously would have different nutritional requirements. Diverse food products are obtained through multiple cropping. Multiple cropping narrows the space available for weeds to grow and hamper their growth through exudation of allelochemicals. Multicropping has a number of environmental benefits, including: i) Increased yields: monocropping has been shown to decrease yields over time, while multiculture practices such as crop rotation and the use of cover crops can increase yields by reducing pests, improving soil health, and increasing water retention. ii) Decreased pest susceptibility: multicropping reduces extreme vulnerability to a wide array of pests, including weeds, insects, fungi, and other organisms. The lack of genetic diversity on monocropped farms means that a single pest can decimate large areas of cropland. iii) Increased biodiversity: multicropped farms have a number of species that may interact in a useful way, such as providing shade for other crops, providing nitrogen fixation for the soil or repelling pests.

Minimum or No Pesticide Use

Various agricultural chemicals are used to kill weeds (herbicides), insects (insecticides), bacteria (microbicides) and fungi (fungicides). Industrial crop production heavily depends upon pesticides because the practice of monocropping increases sensitivity to pests. Unfortunately, pesticides can cause health problems to the workers in farm who apply the chemicals and who participated in various operations in the farm and consumers who eat foods with pesticide residues. Various pesticides have been linked to different kinds of severe health problems. Pesticides also cause environmental damage such as water pollution and soil contamination. In addition, pesticides have been shown to cause declines in pollinators and other beneficial insects that are critical to the health of agricultural systems.

In sustainable agriculture, pesticide use is greatly reduced by the following measures: i) integrated pest management: integrated pest management is a pest-management system that integrates several pest-management approaches. ii) intercropping and companion planting: intercropping is a method of planting crops in closer spacing in order to suppress weeds, to encourage plant diversity in order to avoid insect and pest infestation. Companion planting is a related method that depends on plants that are natural pest repellants (e.g., marigolds), plants that are more attractive to pests than the primary crop or plants that attract beneficial insects. iii) use of beneficial organisms: there are a number of beneficial insects and organisms that, when released, destroy harmful pests. Beneficial insects include predators such as ladybugs; beneficial organisms include nematodes (microscopic worms) that are used to destroy the larvae of pests.

Soil Health Management

Soil health management is a critical component of sustainable agriculture and comprises a number of different growing practices and principles. There is also some evidence that sustainably grown plants may be higher in vital macro and micronutrients, resulting from increased soil health as a direct consequence of organic growing methods and sustainable practices. Some of these practices are organic fertiliser use, crop rotation, intercropping and mulching.

Conclusion

Sustainable agriculture is often summarised by three main goals: economic profitability, environmental stewardship and social responsibility. What may be sustainable at one farm or community may not be sustainable at another. As others have stated, sustainable agriculture is a journey rather than a destination. Making the transition to sustainable agriculture is an essential process. For farmers, the transition to sustainable agriculture normally requires a series of small, realistic steps. Family economics and personal goals influence how fast or how far participants can go in the transition. It is important to realise that each small decision can make a difference and contribute to advancing the sustainable agriculture.

References

A Study of Sociocultural Value of Family Farm. www.akgdev.com/wfnc/resources/Thesis

Center for Integrated Agricultural Systems. *www.cias.wisc.edu.*

National Sustainable Agriculture Information Services (NSAIS). *www.attra.ncat.org*

Sustainable Agriculture Research and Education Program, Agriculture Sustainability Institute, UC Davis. *www.sarep.ucdavis.edu.*

University of Kentucky, College of Agriculture, Food and Environment. *www.uky.edu.*

Section 4
Sustainable Energy and Livelihoods

11

OMANJANA GOSWAMIA, SUMIT MISHRA,
GAUTAM SEN and USHA JHA

A Study of the Resource Recovery Mechanism of the East Kolkata Wetlands Ecosystem

Introduction

The East Kolkata Wetlands (EKW) is a unique example of integrated resource recovery system through productive waste-use activities. The wetlands are located on the eastern fringes of the Kolkata metropolis, covering an area of 12,500 ha. For numerous decades, various groups of local fishermen, farmers and scavengers have been supporting their livelihood by using wastewater. The recovery system is the largest in the world and popularly known as kidney of the city. Wastewater flows through the fishpond covering 4,000 ha which facilitate various physico-chemical processes that improve the quality of water. In 2002, by virtue of its unique resource recovery activities, the Ramsar Bureau declared the EKW as a Ramsar site making it a "Wetland of International Importance". Nearly 10,000 tonnes of fish is produced annually from the 254 *bheris* or fish ponds presently in the EKW area; 550,000 m^3 of wastewater per day from Kolkata which flows through the network of canals is used for irrigation and aquaculture practices. The rest of the water is discharged into the Bay of Bengal. The total EKW area can be divided into four types of land use classes: water-body-oriented area (47%), agricultural area (38%), garbage/productive farming area (5%) and settlement area (10%). The major share of the land is water inundated area and unfit for productive use.

Bheris are 50-150 cm deep flat bottomed fish ponds. The shallow depth allows infiltration of sunlight and constant aeration. The wetland waste stabilisation pond preparation is one of the essential steps of the water purification in the winter months (between November and March) and the pond is subjected to sun air drying. The bottom is then ploughed and mixed with lime and left untouched for a month. The raw sewage is introduced from sewage channels and filled up to a depth of 90 cm. The colour at the time of input is deep black within 3-4 days of standing time, the colour changes to green. This is an indication of algal growth. Within 7 days, algal growth reaches its peak and eutrophication sets in. The algae are removed by the fishermen through their fishing net, rendering the *bheri* water crystal clear. The water is allowed to stand and therefore used for aquaculture comprising pisciculture and agricultural irrigation.

At day 0, the sewage is deep black in colour with foams floating on surface. At day 7, colour becomes green due to intense algal bloom. At day 10, the water becomes transparent and colourless indicating considerable improvement in water quality.

Various studies have been conducted in last 20 years to determine the sewage quality and water purification mechanism. The quality of sewage is determined by assessment of some important properties such as BOD (biochemical oxygen demand), COD (chemical oxygen demand), etc. This arises because of the complex interaction between the physico-chemical and biotic components present together. Factors do not act independently but conglomerates to give a cumulative interaction. Purification of wastewater is an ecology-based technique and studies show that the efficiency in reducing BOD of waste is above 80 per cent and average reduction of coliform bacteria is 99.99 per cent. The solar radiation received is about 250 Langley's per day and enough for adequate photosynthesis. The *bheri* acts as solar reaction, energy being trapped by plankton population. Fishes survive on consumption of plankton. Degradation of organic matter in wastewater brought about plankton. Fishes play the watchdog role by grazing on plankton to prevent overgrowth and also convert available nutrients into readily consumable forms. To determine the purification mechanism, it is critical to know the interactions and relationships amongst the components of a pond system in EKW and their influence on the purification process.

This study looks into the process of nutrient recovery in a single pond of the EKW. The elemental analysis has been carried out to determine the toxicity limits of the edible products obtained as a part of the natural resource recovery system. This study focuses particularly on determination of heavy metal concentrations (lead, cadmium and chromium) in various components of the wetland ecosystem *viz.*, fish, water hyacinth (*Eichhornia crassipes*), soil and water.

Methods

The Study Area

The wetland area lies between latitudes 22° 25″ to 22° 40″N and longitude 88° 20″ to 88° 35″E, and the region is suitable for solar radiation. Climate is more or less sub-tropical with the annual mean rainfall around 200 cm. The maximum temperature during summer rises to 39°C while minimum temperature during winter is 10°C.

Collection of Water Samples

Water samples were collected from four different points of the same pond and were kept in pre-washed sterile containers. Since in the middle of the day ecosystem becomes most active, all the samples for this study were collected in the middle of the day.

Analysis of Water Samples

In this study, different parameters were analysed to monitor the quality of water and all the analyses were commenced on the same day the samples were collected. The non-conservative parameters like pH and dissolved oxygen (DO) were analysed onsite during the time of collection by an automatic portable analyser (WTW Multi 340i). For the analysis of BOD, COD, total nitrogen, NO3-N, NO2-N and PO4-P, water samples were brought to the laboratory in freezing condition and analysed by standard methods as outlined in American Public Health Association (1998).

Analysis of Fish Samples

Fish samples were collected from the pond under study and preserved under freezing condition. Each fish was dissected and then the liver, gill and muscle were taken out. Acid digestion (in nitric acid) was carried out with the samples in separate round bottom flasks.

Results and Discussion

The physico-chemical parameters reveal that the water is slightly alkaline with the pH value ranging from 7.0 to 7.2. The data show that sampling was done much after the release of wastewater because the pond characteristics are almost stable. BOD is on the lower side indicating that wastewater purification process has almost been completed.

Table 11.1

Physico-chemical Parameters

Spot	*pH*	*Temp (°C)*	*Ec (µS/cm)*	*Salinity (ppt)*	*BOD (mg/L)*	*COD (mg/L)*	*DO (mg/L)*	*Alkalinity (mg/L)*	*Total Hardness (mg/mL)*
1	7.28	28	772	0.1	4.95	67.2	4.416	125	135
2	7.19	28	779	0.1	5.25	67.2	4.968	150	145
3	7.09	28	831	0.2	5.4	67.2	3.312	200	160
4	7.14	28	804	0.2	6.15	67.2	5.152	175	155

Spot	*NO3- N (µg/L)*	*NO2-N (µg/L)*	*Total N (%)*	*Total P (%)*
1	166.0	248.1	0.3	1.7
2	138	46.6	0.5	1.5
3	212.5	30.4	0.4	2.5
4	113.6	154.0	0.4	1.7

The samples were analysed for the heavy metals Lead (Pb), Chromium (Cr) and Cadmium (Cd). Heavy metal content was determined in the soil, water, water hyacinth and fish sample taken from the pond under study. Analysis was done using atomic absorption spectrophotometer after the sample preparation. Tabulated results as well as their corresponding graphical representations are presented below.

Table 11.2

Heavy Metal in Soil Sample

Pb(mg/kg)	*Cr(mg/kg)*	*Cd(mg/kg)*
29.6875	12.5375	2.925

Pb is the predominant heavy metal found in the soil samples. It is consistent with observations from other components of the ecosystem, where Pb is found to be the chief heavy metal contaminant.

Figure 11.1

Heavy Metal Concentration in Soil Sample

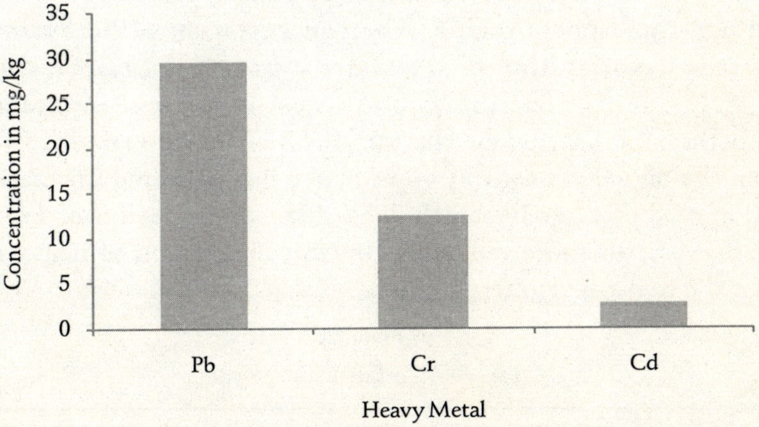

Table 11.3

Heavy Metal in Water Hyacinth

Sampling Point	*Pb (mg/kg)*	*Cr (mg/kg)*	*Cd (mg/kg)*
Rhizosphere	42.91	14.25	4.73
Non-rhizosphere	23.82	0	1.34

Figure 11.2

Heavy Metal Concentration in Water Hyacinth

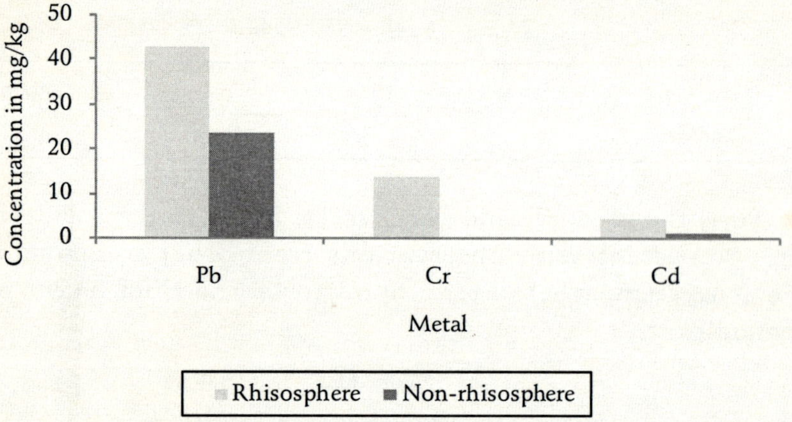

The rhizosphere region shows more heavy metal content than the non-rhizosphere region. The concentration of heavy metal in water is less than that of the water hyacinth (*Eichhornia crassipes*) possibly due to rhisofilteration process occurring simultaneously with the stabilisation of the pond. This can be explained further from the higher concentration of heavy metals found in rhizosphere region of the water hyacinth. The rhizosphere region of *Eichhornia crassipes* shows more tendency towards absorption of heavy metals than the non-rhizosphere region.

Table 11.4

Heavy Metal in Water Sample

Spot	*Pb (mg/kg)*	*Cd (mg/kg)*	*Cr (mg/kg)*
1.	0.0260	0.0025	0.0034
2.	0.0251	0.0026	0.0040
3.	0.0305	0.0030	0.0043
4.	0.0299	0.0026	0.004

Figure 11.3

Heavy Metal Concentration in Water Sample

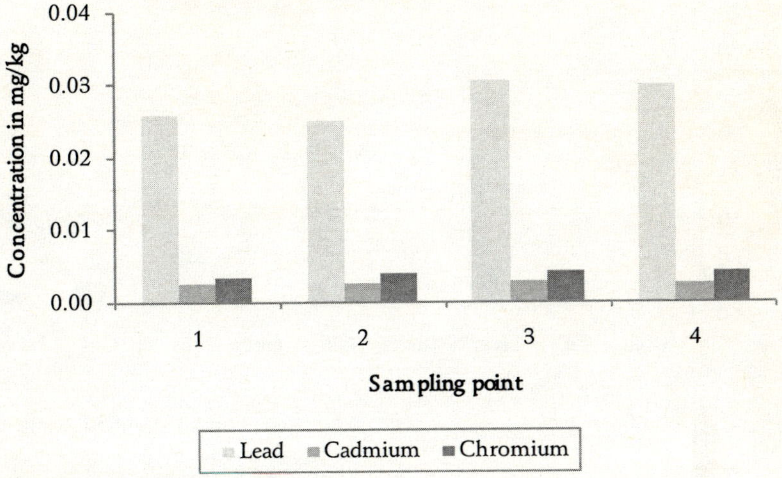

Pb remains the metal present in highest concentration at all points of sampling.

Table 11.5

Heavy Metal in Fish

Name of Fish	*Organs Studied*	*Pb (mg/kg)*	*Cd (mg/kg)*	*Cr (mg/kg)*
Labeo rohita	Muscle	13.160	0.790	2.030
	Gill	8.903	0.947	4.115
	Liver	13.533	4.867	18.333
Tilapia nilotica	Muscle	4.115	0.385	2.022
	Gill	2.361	0.926	2.750
	Liver	5.927	0.681	1.134
Cirrhinus mrigala	Muscle	14.525	6.225	0.324
	Gill	7.543	7.033	2.883
	Liver	31.318	1.758	2.000

Figure 11.4

Heavy Metal Distribution in Fish Organs

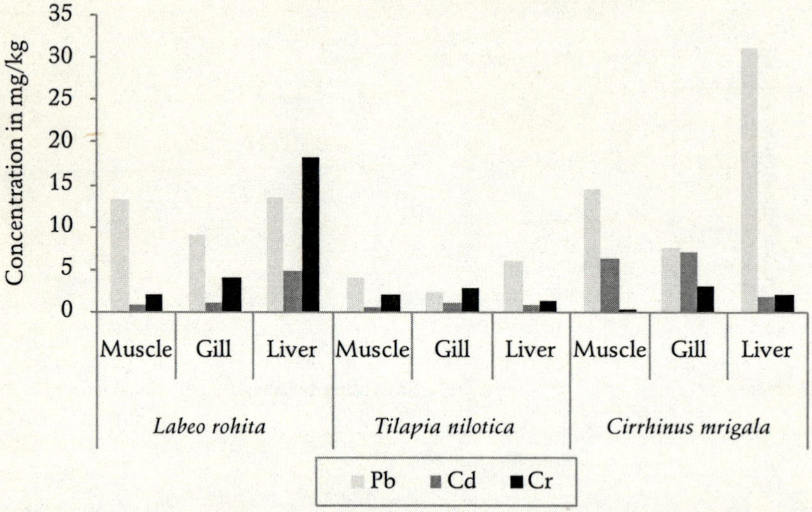

Amongst all the metals studied, Pb is found to be significantly higher than the other two metals, followed by Cr and Cd. Liver seems to be the integral organ for bio-accumulation as can be seen in Figure 11.4 followed by muscle and gill. However, it is alarming that heavy metal is found in significant amount in muscles. Cr is also found in relatively higher concentration especially in *Labeo rohita,* while *Tilapia nilotica* shows the least concentration of heavy metals amongst all fishes. This can be attributed to the fact that *Tilapia nilotica* is a bottom feeder, generally living near the bottom of the fish pond.

In a system that receives heavy metal inputs from various sources, it can be feared that bio-accumulation might occur. The concentration of heavy metal shows a decrease from the soil to the aquaculture product, i.e., fish. Thus, bio-accumulation does not occur, rendering the fish produce safely edible.

Conclusion

All the elements mentioned in the discussion have distinctive roles for the normal physiology of the body. The concentrations of the heavy metals present in a single pond system are well below the prescribed toxicity limits. Fish being the major aquaculture product of the wetlands ecosystem has an annual produce of 10,000 tonnes. It is thus fit for consumption. Hence the resource recovery process practised at EKW is safe, reliable and of standard quality. The entire detoxification process is free from any type of chemical treatment and is purely driven by natural mechanisms and the products generated by the integrated resource recovery mechanism are of consumable quality. The resource recovery system at EKW leads to significant cost savings and nurtures the ecosystem of the entire region. Hence, it can be asserted that this mechanism is safe from the physiological viewpoint and also productive from the economic and environmental perspectives.

References

American Public Health Association (1998). *Standard Methods for the Examination of Water and Wastewater* (20th edition). Baltimore, Maryland: APHA.

Bhupander Kumara, K.S. (2010). "Distribution, Partitioning, Bioaccumulation of Trace Elements in Water, Sediment and Fish from Sewage Fed Fishery Ponds of Kolkata", *Toxicological & Environmental Chemistry*: 243-60.

Chattopadhyay, B., A. Chatterjee and S. K.Mukhopadhyay (2002). "Bioaccumulation of Metals in the East Calcutta Wetland Ecosystem", *Aquatic Ecosystem Health and Management* 5: 191-203.

East Kolkata Wetland Management Authority (2011). *Conservation and Management Plans of East Kolkata Wetlands, India*. Report to be submitted to the Ministry of Environment and Forests, Government of India.

Furedy, C. and D. Ghosh (1984). "Resource Conserving Traditions and Waste Disposal: The Garbage Farming and Sewage-fed Fisheries of Calcutta", *Conserv. Recycling* 7(2–4): 159-65.

Ghosh, A.K. (1990). "Biological Resources of Wetlands of East Kolkata", *Indian Journal of Landscape System and Ecological Studies* 13: 10-23.

Ghosh, D. (1983). *Sewage Treatment Fisheries in East Calcutta Wetlands*. Report to the Department of Fisheries, Government of West Bengal, Calcutta, India.

Ghosh, D. and S. Sen (1987). "Ecological History of Calcutta's Wetland Conservation", *Environ. Conserv*. 14(3): 219-26.

Kundu, N., M. Pal and S. Saha (2008). "East Kolkata Wetlands: A Resource Recovery System through Productive Activities", in M. Sengupta and R. Dalwani (eds.), *The 12th World Lake Conference* (Taal), 28 October-2 November, Jaipur, India, pp.868-81.

Mukherjee, D.P. (2011). "Stress of Urban Pollution on Largest Natural Wetland Ecosystem in East Kolkata-causes, Consequences and Improvement", *Archives of Applied Science Research*: 443-61.

Raychaudhuri, S., M. Mishra, S. Salodkar, M. Sudarshan, and A. R.Thakur (2008). "Traditional Aquaculture Practice at East Calcutta Wetland: The Safety Assessment", *American Journal of Environmental Sciences* 4: 173-7.

Saha, T. and Ghosh, P.B. (2003). "Wise Use of Wetland in Water Conservation and Advance Irrigation Methods—A Case Study", in technical volume of the *National Seminar on Role of Agricultural Engineers in Water Conservation*. India: The Institute of Engineers. pp. 74-8.

Shaon Raychaudhuri, S.S. (2002). "Integrated Resource Recovery at East Calcutta Wetland: How Safe is These?" *American Journal of Agricultural and Biological Science* 2: 75-80.

Acknowledgement: The authors would like to thank the Institute of Environment Studies and Wetland Management (IESWM) for giving the permission to conduct the study and work in the laboratory. They would like to acknowledge the Principal Scientist of IESWM Mr Tapan Saha for extending his cooperation at all stages of conducting the study and permitting to use the results.

12

PRAJNAPARAMITA MISHRA and
M. GOPINATH REDDY

Improving Livelihoods or Intensifying Poverty

Coal Mining in Chhattisgarh

Introduction

In India's energy scenario, coal is the most dominant energy source meeting around 52 per cent of primary commercial energy needs. India is the third largest coal-producing country in the world after China and the US (Coal India Ltd. [CIL] 2013). According to the Geological Survey of India, the country has 2,93,497 million tonnes of geological resources of coal. Jharkhand has the highest reserves (27.38%), followed by Odisha (24.34%) and Chhattisgarh (17.32). CIL as an organised state-owned coal mining corporate is the single largest coal producer in the world, producing around 81.1 per cent of India's overall coal production. The Government of India (GoI) has permitted CIL to select state-owned enterprises. Therefore, CIL has subsidiaries working in all the states of India having coal reserves.

Indian literatures show that livelihoods of people in coal mining areas of India are much better than non-coal mining areas (Lahiri-Dutt et al. 2012, Bhusan and Hazra 2008). This is because public sector coal mining companies have employed more people. The coal-producing districts statistics also show that they rank higher in Human Development Index (HDI) and has a lower poverty ratio compared to the state's average. With this background, our present study tried to look at the impacts of coal mining on livelihoods in the coal-rich area of Chhattisgarh. Our studies try to find out does this richer means rich in all aspect? Do they have livelihoods which are sustainable?

The present paper has been divided as follows. Section one is the introduction to the study. Section two discusses about general mining in Chhattisgarh and coal mining in particular. The next section three introduces the study area and presents the sampling technique. Section four presents the result of this study. Finally, section five concludes the study.

Coal Mining in Chhattisgarh

Chhattisgarh, a state in central India is very rich in mineral reserves. Jharkhand, Odisha and Chhattisgarh have almost all the coal deposits of India. The state has also all the tin ore. It houses the best quality iron ore deposits in the world, Bailadila mines in South Chhattisgarh. The state has also rich deposits of bauxite, limestone, Dolomite and corundum. Deposits of Diamond, gold and Alexandrite, one of the rarest gemstones, are also found in the state (Chhatisgarh Mineral Development Corporation [CMDC] 2013).

CMDC undertakes scientific exploration, commercial exploitation and trading of minerals in the state. According to CMDC, they undertake partnerships with national and international private sector mineral companies so that natural wealth of the state can be translated more efficiently into prosperity of its people (CMDC 2013).

The coal deposits of Chhattisgarh and Madhya Pradesh comes under South Eastern Coalfields Limited (SECL). It is the largest coal-producing company in the country. It is one of the eight subsidiaries of CIL (a GoI undertaking) under the Ministry of Coal. In the year 2012-13, total coal production by SECL was 118.33 million tonnes out of total coal production of 452.211 million tonnes produced by CIL. This is highest among all subsidiaries of CIL, and among all coal-producing companies in India, SECL is making profits since its inception (SECL 2013).

The coal deposits of SECL occur in five districts of Bilaspur, Korba, Raigarh, Surguja and Korea in Chhattisgarh and three districts of Shahdol, Umaria and Anuppur in Madhya Pradesh in the Son-Mahanadi master basin. SECL has 92 mines. Total UG (underground) mines are 70 and total OC (open-cast) mines are

21. There is 1 mixed mine. Out of that 42 UG mines, 13 OC mines and 1 mixed mines are in Chhattisgarh. SECL corporate office is at Bilaspur. SECL has four major coalfields. Korba coalfields in Korba districts, Central coalfields in Surguja and Korea districts, Mand-Raigarh coalfields in Raigarh district and Ramkola-Tatapani coalfields in Surguja district (SECL 2013).

Study Area and Sample Selection

Both primary and secondary sources of information have been collected for this study. For primary data collection, the survey was carried out in different phases from October 2012 to May 2013. Our study areas are situated in the coal-bearing areas of Chhattisgarh. The survey covered 300 households. Coal deposits of Chhattisgarh come under SECL. The coal deposits of SECL occur in five districts of Chhattisgarh i.e., Bilaspur, Korba, Raigarh, Surguja and Korea. Out of this, we have taken three districts i.e., Korea, Surguja and Surajpur (on 1 January 2012, this district was carved out from Surguja). Based on the secondary data, the research team prepared a list of mine sites in these three districts. The idea was to take new mining area between 5 and 15 years age mines, if not available then to go for older mines.

In the Korea district of Chhattisgarh, mines are operational in three blocks (Table 12.1). Out of these three, Manendragarh and Khadgawn are selected. In the Manendragarh block, out of four mines area, two are in operational form. Out of that two, Palkimada mine area was selected. This area has one UG mine and it also borders an OC mines of Madhya Pradesh. Therefore, the surrounding villagers are facing the impacts of both mines. In Khadgawn block, Chirimiri Arrow mine area is selected as it is an OC mine and not very old.

In the Surajpur district of Chhattisgarh, mines are operational in three blocks (Table 12.2). Out of these three, Srinagar and Premnagar are selected. In the Srinagar block, Amgaon is the only mines area (surrounding villages under two *panchayats*) and was taken. In Premnagar block, Rehar and Gayatri areas were taken (Ketki is only five years old).

Table 12.1

Coal Mines in Korea District, Chhattisgarh

District	Block	Panchayat	Mines Area	Type of Mines					Remark
				UG	Age of Mine	OC	Age of Mine	Total	
Korea	Manendragarh	Lediri	BSEAM Colery	1	40 yrs	-	-	1	Closed
		Jhagrakhand	Jhagrakhand	1	40 yrs	-	-	1	Closed
			West Jhagrakhand	1	30 to 40 yrs	-	-	1	Running
		Khongapani	Palkimada	1	30 to 40 yrs	-	-	1	Running
	Baikunthpur	Churcha	Churcha	1	30 to 40 yrs	-	-	1	Running
		Khond	Jilimiri	1	20 to 30 yrs	-	-	1	Running
			Pondapora	1	20 to 30 yrs	-	-	1	Running
		Katkona	Katkona	1	30 to 40 yrs	-	-	1	Running
	Khadgawn	Chirimiri	Kurasia	1	30 to 40 yrs	-	-	1	Running
			Haldiwada	1	30 to 40 yrs	-	-	1	Running
			NCPH	1	30 to 40 yrs	-	-	1	Running
			Chirimiri Arrow	-	-	2	10 -15 Yrs	2	One is running and other one is closed
			Chilawada	1	30 to 40 yrs	-	-	1	Running
		Total		12	-	2	-	14	Out of 14 mines, 3 mine have closed and 11 mines are running

Source: Field Study (2012).

Table 12.2

Coal Mines in Surajpur District, Chhattisgarh

District	Block	Panchayat	Mines Area	Type of Mines					Remark
				UG	Age of Mine	OC	Age of Mine	Total	
Surajpur	Srinagar	Salhi	Amgaon	-	-	1	7 yrs (2005)	1	Running
		Amgaon							
	Surajpur	-	Bisrampur	-	-	1	45 yrs	1	Running
			Kumda	1	35 yrs	-	-	1	Running
			Balarampur	1	35 yrs	-	-	1	Running
	Premnagar	Mani	Rehar	1	15 yrs	-	-	1	Running
			Gayatri	1	15 yrs	-	-	1	Running
			Ketki	1	5 yrs	-	-	1	Running
	Total			5	-	2	-	7	All 7 mines are running

Source: Field Study (2012).

In the Sarguja district of Chhattisgarh, mines are operational in only one block (Table 12.3). It has only two mines area and both are selected. To summarise, in Chhattisgarh, six mines area were selected (three OC and three UG).

Table 12.3

Coal Mines in Sarguja District, Chhattisgarh

District	*Block*	*Panchayat*	*Mines Area*	*Type of Mines*					*Remark*
				UG	*Age of Mine*	*OC*	*Age of Mine*	*Total*	
Sarguja	Lakhanpur	Getwa	Gayatri[1]	1	15 yrs	-	-	1	UG is running
		Amera	Amera	-	-	1	5 yrs	1	OC is running
		Total		1	-	1	-	2	Both are working

Note: 1. Gayatri UG mines fall under both Surajpur and Sarguja Districts.

Source: Field Study (2012).

Out of these three districts of Chhattisgarh, 100 households are selected from each district making it 300 in total (Table 12.4).

Coal Mining, Environment and Livelihoods in Chhattisgarh

According to Joshi et al. (2006), mining causes to main environmental problems in Chhattisgarh: one, pollution of rivers and streams and two, alluvial erosion and deforestation. Their study shows that forest patches near Korba coal mining area have been badly degraded, whereas patches away from the mining sites registered less impact. According to them, around 78.49 per cent forests have been affected because of mining. Around 5.93 per cent have been totally converted to industrial set-up, 55.31 per cent showed medium changes attributed to formation of barren and wasteland, and around 17.25 per cent has been degraded. Beg et al. (2011) study shows incidence of high fluoride in groundwater in parts of Raigarh district.

Table 12.4

Details of Sample Households, Chhattisgarh

District	Mines Area	Villages	Total Wards	Sample Wards	Total Households					Sample Households				
					SC	ST	OBC	OC	Total	SC	ST	OBC	OC	Total
Korea	Palkimada	Kongapani	15	2	274	300	4	4	582	23	23	2	2	50
	Chirimiri Arrow	Chirimiri	40	2	31	745	101	30	907	6	28	10	6	50
Surajpur	Amgaon	Salhi	3	3	0	198	78	0	276	1	39	10	0	50
	Rehar	Podi	4	4	7	200	86	7	300	2	16	7	0	25
	Gayatri	Mani	1	1	0	180	20	0	200	0	16	9	0	25
Sarguja	Gayatri	Getwa	8	8	15	366	19	0	400	11	35	4	0	50
	Amera	Amera	1	1	2	98	20	0	120	0	35	15	0	50
Total			72	21	329	2087	328	41	2785	43	192	57	8	300

Scurce: Field Study (2012).

Table 12.5

Total Income from Various Sources

District	*Village*	*Area*	*Agriculture*	*Wage Labour*	*Coal Mines Employ*	*Coal Mines Contract/Wage Labour*	*Others*	*Total*
Korea	Kongapani	Kongapani : UG Mines	3600 (0.1)	1564400 (31.2)	3052000 (60.9)	0 (0.0)	389300 (7.8)	5009300 (100.0)
	Chirimiri	Chirimiri : OC and UG Mines	423600 (5.4)	526800 (6.7)	5183400 (65.8)	0 (0.0)	1738500 (22.1)	7872300 (100.0)
Surajpur	Salhi	Salhi : Amgaom OC Mines	1012800 (9.3)	1274500 (11.8)	7704000 (71.1)	28600 (0.3)	822800 (7.6)	10842700 (100.0)
	Podi	Gayatri : UG Mines	489200 (17.1)	516200 (18.0)	480000 (16.8)	866800 (30.2)	513400 (17.9)	2865600 (100.0)
	Mani	Rehar : UG Mines	672500 (33.7)	647000 (32.4)	0 (0.0)	150000 (7.5)	526600 (26.4)	1996100 (100.0)
Sarguja	Getra	Rehar : UG Mines	604000 (9.0)	1256000 (18.8)	3323000 (49.7)	442000 (6.6)	1061500 (15.9)	6686500 (100.0)
	Amera	Amera : OC Mines	654800 (9.1)	947000 (13.2)	5236000 (72.9)	180000 (2.5)	168700 (2.3)	7186500 (100.0)
	Total		3860500 (9.1)	6731900 (15.9)	24978400 (58.8)	1667400 (3.9)	5220800 (12.3)	42459000 (100.0)

Source: Field data (2012).

Diversified Rural Livelihoods

In the district of Korea, in both the study sites, 24 per cent of the households have employment in coal mines. The main differences in employment in both these areas are in agriculture and wage labour. In the Kongapani area, percentage of wage labour is very high (66%). After mining, there was migration of outsiders to this area as there was an opportunity for work. As agricultural lands are taken for mining, percentage of agriculture is less (10%). However, in Chirimiri area still 34 per cent of households depend on agriculture. Here mining is mainly going on in reserve forest area. Therefore, agricultural lands are not taken. That also resulted in less percentage of wage labour (22%).

In Surajpur, Salhi-Amgaom OC mines, nearly half of the households are employee in mines (48%). Still 36 per cent of households are dependent on agriculture. Here people are cultivating their land, which is already acquired by the mining company. They may lose these lands in future. That also results in a less percentage (10%) of wage labour.

In Podi-Gayatri UG mines area, agriculture still plays an important role as this is an underground mine (68%). Here percentage of coal mines employee are less (4%) and percentage of contract wage labour is high (20%). As their lands are not acquired, there is no permanent employment in the mines. Same is the case of Mani.

In Getra-Rehar UG mines area, half of the households are employed in agriculture (50%). Here also lands are not acquired as this is an UG mine. For some households whose lands are acquired, they have a job in the mines (22%). Wage labour has a percentage of 14 per cent and 4 per cent are contract employee. In Amera area, lands are acquired but agriculture is going on (52%). That's why percentage of mining employee is also high (42%).

The above explanation is also reflected in Table 12.6. Percentage of female workers is very less in the mining villages. The only place where they form a small part of the workforce is where agriculture is going on. In these eight villages, only six women have job in the coal mines.

Table 12.6

Educational Status of the Sample Households

Educational Status	Korea				Surajpur						Surguja			
	Kongapani		Chirimiri		Salhi		Podi		Mani		Getra		Amera	
	Kongapani OC and UG Mines		Chirimiri OC and UG Mines		Salhi-Am gaom OC Mines		Gayatri UG Mines		Rehar UG Mines		Rehar UG Mines		Amera OC Mines	
	Male	Female	Male	Female	Male	Female	Male	Female	Male	Female	Male	Female	Male	Female
Illiterate[1,2]	12 (9.8)	24 (22.2)	28 (20.6)	39 (32.2)	23 (17.2)	55 (42.0)	15 (20.3)	27 (45.8)	6 (10.7)	19 (35.2)	26 (20.6)	40 (34.8)	28 (23.0)	55 (47.8)
Literate (non-formal)	10 (8.2)	23 (21.3)	10 (7.4)	7 (5.8)	4 (3.0)	3 (2.3)	9 (12.2)	1 (1.7)	3 (5.4)	2 (3.7)	5 (4.0)	6 (5.2)	8 (6.6)	4 (3.5)
Literate below primary	1 (0.8)	1 (0.9)	2 (1.5)	7 (5.8)	5 (3.7)	6 (4.6)	8 (10.8)	4 (6.8)	2 (3.6)	1 (1.9)	10 (7.9)	5 (4.3)	0 (0.0)	1 (0.9)
Primary	36 (29.5)	19 (17.6)	23 (16.9)	16 (13.2)	29 (21.6)	33 (25.2)	17 (23.0)	12 (20.3)	12 (21.4)	11 (20.4)	36 (28.6)	22 (19.1)	26 (21.3)	21 (18.3)
Middle	23 (18.9)	23 (21.3)	31 (22.8)	20 (16.5)	36 (26.9)	20 (15.3)	15 (20.3)	8 (13.6)	15 (26.8)	12 (22.2)	20 15.9)	26 22.6)	32 26.2)	19 16.5)
Secondary	16 (13.1)	8 (7.4)	24 (17.6)	18 (14.9)	15 (11.2)	7 (5.3)	4 (5.4)	3 (5.1)	8 (14.3)	5 (9.3)	15 (11.9)	9 (7.8)	7 (5.7)	7 (6.1)
Inter (10+2)	19 (15.6)	8 (7.4)	14 (10.3)	12 (9.9)	21 (15.7)	7 (5.3)	5 (6.8)	4 (6.8)	7 (12.5)	2 (3.7)	10 (7.9)	5 (4.3)	19 (15.6)	7 (6.1)
Graduation and above	5 (4.1)	2 (1.9)	4 (2.9)	2 (1.7)	1 (0.7)	0 (0.0)	1 (1.4)	0 (0.0)	3 (5.4)	2 (3.7)	4 (3.2)	2 (1.7)	2 (1.6)	1 (0.9)
Total	122 (100.0)	108 (100.0)	136 (100.0)	121 (100.0)	134 (100.0)	131 (100.0)	74 (100.0)	59 (100.0)	56 (100.0)	54 (100.0)	126 (100.0)	115 (100.0)	122 (100.0)	115 (100.0)

Note: 1. The working definition of literacy in the Indian census since 1991 is: the total percentage of the population of an area, at a particular time, aged seven years or above, who can read or write with understanding. Here the denominator is the population aged seven years or more. The same criterion has been retained in the 2011 Census. This study considers this definition.

2. According to the 2011 Census, the literacy rate of Korea District is 70.64 per cent, with a male literacy rate of 80.37 per cent and female literacy rate of 60.60 per cent. Surajpur has a literacy rate of 57.79 per cent. The literacy rate of Surguja District is 60.01 per cent, with a male literacy rate of 96.53 per cent and female literacy rate of 50.32 per cent. However, Chhattisgarh has a literacy rate of 70.28 per cent, with a male literacy rate of 80.27 per cent and female literacy rate of 59.58 per cent.

Source: Field data (2012).

Table 12.7

Land Holding in Sample Households (%)

District	*Village*	*Area*	*Landless*	*Marginal Farmer*	*Small Farmer*	*Medium Farmer*	*Large Farmer*	*Total*
Korea	Kongapani	Kongapani : OC and UG Mines	48 (96.0)	2 (4.0)	0 (0.0)	0 (0.0)	0 (0.0)	50 (100.0)
	Chirimiri	Chirimiri : OC and UG Mines	36 (72.0)	4 (8.0)	5 (10.0)	4 (8.0)	1 (2.0)	50 (100.0)
Surajpur	Salhi	Salhi : Amgaom OC Mines	8 (16.0)	19 (38.0)	14 (28.0)	6 (12.0)	3 (6.0)	50 (100.0)
	Podi	Gayatri : UG Mines	3 (12.0)	12 (48.0)	4 (16.0)	4 (16.0)	2 (8.0)	25 (100.0)
	Mani	Rehar : UG Mines	4 (16.0)	6 (24.0)	11 (44.0)	3 (12.0)	1 (4.0)	25 (100.0)
Sarguja	Getra	Rehar : UG Mines	17 (34.0)	15 (30.0)	9 (18.0)	6 (12.0)	3 (6.0)	50 (100.0)
	Amera	Amera : OC Mines	13 (26.0)	28 (56.0)	5 (10.0)	3 (6.0)	1 (2.0)	50 (100.0)
	Total		129 (43.0)	86 (28.7)	48 (16.0)	26 (8.7)	11 (3.7)	300 (100.0)

Source: Field data (2012).

The literacy rate differs across villages in the mining area. The illiteracy rate among women is very high in Amera OC mines area (47.8%) and reasonably good in Kongapani area (22.2%). Among men, it is high in Amera area (23%) and low in Kongapani area (9.8%). The percentage of those educated above graduation is very low among men and women both.

In the district of Korea, percentages of landless households are very high as they are OC mines and agricultural lands are acquired. In Surajpur District, majority households are marginal and small farmers. Comparatively, Sarguja has a higher percentage of landless. All the sample households own a house. In all the villages, houses are thatched or semi-*pucca*. Percentages of *pucca* households are very less. Numbers of rooms are two or more than two.

Apart from house, the households also have other physical assets as given in Table 12.9.

Table 12.10 on household expenditures shows that more than 60 per cent of the household expenditure goes to food expenses. Other expenses like education, travel, health, recreation, cloth, etc., have a very low share.

Conclusion

The analysis shows that people in the mining-affected area do not have sustainable livelihoods. Although they are rich in some capital, they are lagging behind in other forms of capital. For example, they are very poor in natural capital. However, the sustainable development framework requires people to be rich in all forms of capital. Therefore, coal mining in Chhattisgarh is not leading to a sustainable livelihoods.

Table 12.8

Housing Status in the Study Area

District	Village	Area	Type of House				No. of Rooms		
			Thatched	Semi-Pucca	Pucca	Total	1	2	>2
Korea	Kongapani	Kongapani : OC and UG mines	32 (64.0)	14 (28.0)	4 (8.0)	50 (100.0)	3 (6.0)	19 (38.0)	28 (56.0)
	Chirimiri	Chirimiri : OC and UG mines	27 (54.0)	22 (44.0)	1 (2.0)	50 (100.0)	4 (8.0)	14 (28.0)	32 (64.0)
Surajpur	Salhi	Salhi-Amgaom : OC mines	42 (84.0)	8 (16.0)	0 (0.0)	50 (100.0)	2 (4.0)	9 (18.0)	39 (78.0)
	Podi	Gayatri : UG mines	25 (100.0)	0 (0.0)	0 (0.0)	25 (100.0)	1 (4.0)	6 (24.0)	18 (72.0)
	Mani	Rehar : UG mines	23 (92.0)	1 (4.0)	1 (4.0)	25 (100.0)	0 (0.0)	5 (20.0)	20 (80.0)
Surguja	Getra	Rehar : UG mines	48 (96.0)	2 (4.0)	0 (0.0)	50 (100.0)	2 (4.0)	13 (26.0)	35 (70.0)
	Amera	Amera : OC mines	49 (98.0)	0 (0.0)	1 (2.0)	50 (100.0)	2 (4.0)	19 (38.0)	29 (58.0)
Total			246 (82.0)	47 (15.7)	7 (2.3)	300 (100.0)	14 (4.7)	85 (28.3)	201 (67.0)

Source: Field data (2012).

Table 12.9

Physical Assets

Particulars	*Korea*		*Surajpur*			*Sarguja*		*Total*
	Kongapani	*Chirimiri*	*Salhi*	*Podi*	*Mani*	*Getra*	*Amera*	
	OC and UG Mines	*OC and UG Mines*	*Amgaom OC Mines*	*Gayatri - UG Mines*	*Rehar : UG Mines*	*Rehar : UG Mines*	*OC Mines*	
Cycle	21 (9.72)	27 (12.50)	35 (16.20)	26 (12.04)	22 (10.19)	41 (18.98)	44 (20.37)	216 (100.00)
Radio/ Transistor	0 (0.00)	1 (14.29)	2 (28.57)	0 (0.00)	0 (0.00)	1 (14.29)	3 (42.86)	7 (100.00)
Fan	30 (24.19)	46 (37.10)	17 (13.71)	4 (3.23)	9 (7.26)	10 (8.06)	8 (6.45)	124 (100.00)
Almirah	12 (20.00)	17 (28.33)	10 (16.67)	3 (5.00)	5 (8.33)	9 (15.00)	4 (6.67)	60 (100.00)
TV	15 (17.86)	27 (32.14)	14 (16.67)	4 (4.76)	10 (11.90)	12 (14.29)	2 (2.38)	84 (100.00)
Scooter/ Motorcycle	11 (14.47)	16 (21.05)	19 (25.00)	2 (2.63)	5 (6.58)	12 (15.79)	11 (14.47)	76 (100.00)
Refrigerator	0 (0.00)	5 (38.46)	4 (30.77)	0 (0.00)	1 (7.69)	3 (23.08)	0 (0.00)	13 (100.00)
Car	0 (0.00)	0 (0.00)	2 (66.67)	0 (0.00)	1 (33.33)	0 (0.00)	0 (0.00)	3 (100.00)
Sewing Machine	4 (13.33)	15 (50.00)	0 (0.00)	2 (6.67)	3 (10.00)	4 (13.33)	2 (6.67)	30 (100.00)
Watch/Clock	35 (14.77)	55 (23.21)	40 (16.88)	16 (6.75)	22 (9.28)	27 (11.39)	42 (17.72)	237 (100.00)
Chairs	75 (17.24)	114 (26.21)	74 (17.01)	16 (3.68)	47 (10.80)	51 (11.72)	58 (13.33)	435 (100.00)
Cot	74 (8.51)	178 (20.46)	151 (17.36)	76 (8.74)	86 (9.89)	146 (16.78)	159 (18.28)	870 (100.00)
Mobile	32 (15.31)	57 (27.27)	37 (17.70)	11 (5.26)	17 (8.13)	23 (11.00)	32 (15.31)	209 (100.00)

Source: Field data (2012).

Table 12.10

Household Expenditures

District	*Village*	*Area*	*Food Expenses*	*Education of Child*	*Travel Work Expenses*	*Health Expenses*	*Cloths Expenses*	*Recreation Expenses*	*Other Expenses*	*Total*
Korea	Kongapani	Kongapani : OC and UG Mines	179000 (74.6)	7290 (3.0)	1900 (0.8)	8250 (3.4)	17500 (7.3)	7000 (2.9)	19100 (8.0)	240040 (100.0)
	Chirimiri	Chirimiri : OC and UG Mines	191150 (59.9)	30910 (9.7)	14300 (4.5)	10000 (3.1)	25550 (8.0)	15490 (4.9)	31750 (9.9)	319150 (100.0)
Surajpur	Salhi	Salhi : Amgaom OC Mines	163000 (61.6)	11320 (4.3)	12800 (4.8)	7250 (2.7)	12100 (4.6)	12330 (4.7)	45850 (17.3)	264650 (100.0)
	Podi	Gayatri : UG Mines	64050 (71.2)	2600 (2.9)	4100 (4.6)	2950 (3.3)	4400 (4.9)	3800 (4.2)	8100 (9.0)	90000 (100.0)
	Mani	Rehar : UG Mines	59300 (64.6)	3950 (4.3)	6900 (7.5)	2850 (3.1)	4550 (5.0)	4000 (4.4)	10300 (11.2)	91850 (100.0)
Sarguja	Getra	Rehar : UG Mines	171000 (64.1)	23590 (8.8)	16880 (6.3)	6250 (2.3)	10900 (4.1)	9650 (3.6)	28400 (10.6)	266670 (100.0)
	Amera	Amera : OC Mines	154200 (64.6)	12950 (5.4)	18200 (7.6)	5650 (2.4)	10150 (4.3)	9300 (3.9)	28150 (11.8)	238600 (100.0)
	Total		981700 (65.0)	92610 (6.1)	75080 (5.0)	43200 (2.9)	85150 (5.6)	61570 (4.1)	171650 (11.4)	1510960 (100.0)

Source: Field data (2012).

Table 12.11

Food Security[1]

District	Village	Area	Less than 3 Months	3 to 6 Months	6 to 9 Months	9 to 12 Months	Surplus (Above 12 Months)	Total
Korea	Kongapani	Kongapani : OC and UG Mines	0 (0.0)	3 (6.0)	28 (56.0)	19 (38.0)	0 (0.0)	50 (100.0)
	Chirimiri	Chirimiri : OC and UG Mines	0 (0.00)	1 (2.04)	5 (10.20)	43 (87.76)	0 (0.00)	49 (100.0)
Surajpur	Salhi	Salhi : Amgaom OC Mines	0 (0.0)	0 (0.0)	4 (8.51)	43 (91.49)	0 (0.0)	47 (100.0)
	Podi	Gayatri : UG Mines	0 (0.0)	0 (0.0)	0 (0.00)	25 (100.0)	0 (0.0)	25 (100.0)
	Mani	Rehar : UG Mines	0 (0.0)	0 (0.0)	3 (12.50)	21 (87.50)	0 (0.0)	24 (100.00)
Sarguja	Getra	Rehar : UG Mines	0 (0.0)	0 (0.0)	7 (14.29)	42 (85.71)	0 (0.0)	49 (100.0)
	Amera	Amera : OC Mines	0 (0.0)	0 (0.0)	10 (20.41)	39 (79.59)	0 (0.0)	49 (100.0)
	Total		0 (0.0)	4 (1.37)	57 (19.45)	232 (79.18)	0 (0.0)	293 (100.0)

Note: 1. Food security refers to the availability of food and one's access to it. A household is considered food secure when its occupants do not live in hunger and fear of starvation.

Source: Field data (2012).

Table 12.12

Source of Borrowing (Percentage of Households)

District	Korea		Surajpur			Sarguja		Total
Village	Kongapani	Chirimiri	Salhi	Podi	Mani	Getra	Amera	
	OC and UG Mines	OC and UG Mines	Amgaom OC Mines	Gayatri : UG Mines	Rehar : UG Mines	Rehar : UG Mines	OC Mines	
Bank	2 (18.2)	5 (50.0)	1 (25.0)	2 (100.0)	3 (100.0)	3 (75.0)	1 (100.0)	17 (48.6)
Self-help Groups	0 (0.0)	2 (20.0)	0 (0.0)	0 (0.0)	0 (0.0)	0 (0.0)	0 (0.0)	2 (5.7)
Private Lender	9 (81.8)	3 (30.0)	3 (75.0)	0 (0.0)	0 (0.0)	1 (25.0)	0 (0.0)	16 (45.7)
Total	11 (100.0)	10 (100.0)	4 (100.0)	2 (100.0)	3 (100.0)	4 (100.0)	1 (100)	35 (100.0

Source: Field data (2012).

References

Bhushan, Chandra and Monali Zeya Hazra (2008). *Rich Lands Poor People: Is Sustainable Mining Possible*. New Delhi: Centre for Science and Environment

Chhattisgarh Mineral Development Corporation (2013). *Mining*. Available at *http://www.cg.gov.in/mining/mining1.htm* (accessed on 13 November 2013).

Coal India Limited (2013). *About Us*. Available at *https://www.coalindia.in/en-us/company/aboutus.aspx* (accessed on 13 November 2013).

Joshi, P.K., M. Kumar, N. Midha, Vijayanand and A. Paliwal (2006). "Assessing Areas Deforested by Coal Mining Activities through Satellite Remote Sensing Images and GIS in Parts of Korba, Chhattisgarh", *Journal of Indian Society of Remote Sensing* 34(4): 415-21.

Lahiri-Dutt, Kuntala, Radhika Krishnan and Nesar Ahmed (2012). "Land Acquisition and Dispossession: Private Coal Companies in Jharkhand", *Economic and Political Weekly* XLVII(6): 39-45.

South Eastern Coalfields Limited (2013). *Administrative Setup and Coal Deposits of SECL*. Available at *http://www.secl.gov.in/mcoalfields.htm* (Accessed on 13 November 2013).

13

K.R. SASTRY

Mainstreaming Disaster Risk Reduction

Initiatives of Civil Society Organisations and Fresh Leads

Preamble

The global economic damages from natural disasters in 2010 alone amounted to $123.9 billion, besides 297,000 lives were lost and another 217 million were affected. It is interesting to note that every year natural disasters lead to substantial loss of life and livelihoods, costing an estimated $50 billion in the form of destroyed public and private property and mobilisation of emergency response personnel and equipment in the United States (US) alone (Guha-Sapir et al. 2010).

The situation was much worse in India during 2012-13 alone, as the country suffered heavy losses due to incessant calamities like cyclones in the states adjoining the Bay of Bengal and the unusual cloudburst and resultant floods in Uttarakhand, which left the victims of 'religious tourism' shell shocked. The famous Shivite shrine of Kedarnath was nearly wiped out as it was submerged in the unprecedented flooding that breached a massive water tank situated above the shrine.

As can be surmised, disasters, both natural and man-made, repeatedly wipe out years of hard work and the resources invested in development. A major chunk of budgetary allocations by governments aimed at various developmental schemes was diverted from long-term development strategies to short-term relief measures and for the restoration and reconstruction of devastated areas. The effects of disasters on population and life supporting systems are only too discernible and the society is forced to pay a very heavy price alongside sufferings of the victims of various types. Pithily put, India faces productivity and infrastructural losses on a large scale besides millions of lives on account of natural and human-induced disasters caused by a combination of geographical, ecological, human and developmental factors with regular tenacity.

A one-time disaster may eclipse all the development achieved over time as it takes time to recover from the debilitating effects caused by calamities. Put concisely, development is a planned change with a larger human purpose and so is the case of management of disasters. Development essentially means 'a better quality of life for people.' In the Indian parlance, it actually means 'poverty alleviation,' and for administrators, planners, scholars, professionals, and development pundits, it may mean 'economic growth and prosperity.' On India's developmental landscape, DRR (disaster risk reduction) should have a precise marking, which should help development *vis-à-vis* resilience of the communities to face calamities with equanimity.

Alternatively, the type of development choices made by some countries such as growth in high risk zones creates disaster risks. That said, there appears to be a significant alliance between disasters, development and resilience of communities, and they affect one another. In broad terms, development, DRR and resilience of communities comprise the core concerns of sustainable human development.

Preparedness and Risk Reduction: National Concerns

The Government of India (GoI) came to grips with the problem of fine-tuning the country's disaster management strategies in order to shift the focus from the traditional relief and recovery-oriented aspects to disaster preparedness and mitigation. According to the Prime Minister and Chairman of National Disaster Management Authority (NDMA 2005)[1], the apex body was established to build a safer and disaster resilient India by mounting a holistic, pro-active, multi-disaster and technology-driven strategy for disaster

1. Vision Statement by the Prime Minister on NDMA.

management through collective efforts of all government agencies and NGOs (non-governmental organisations). M. Sashidhar Reddy (NDMA 2005) Vice Chairman noted that NDMA would strive to empower the stakeholders for improving effectiveness of disaster management in India.

Disasters frequently happen while interfacing natural events with the people, livelihoods, economies and infrastructure. Therefore, development planning has become more challenging in light of increased frequency of disasters. The over-emphasis on growth has also contributed to environmental degradation in resource rich as well as resource deficient regions. Also, environmental management has a multitude of concerns with pollution making people sick. The anxieties *vis-à-vis* pollution would affect people's livelihoods and on top the negative practices might increase poverty and distress that may turn into a source for further deprivation.

To be objective, the consequences of growth on demand for forest products would be disastrous unless accompanied by timely re-forestation measures. The high levels of energy consumption and alarming levels of carbon emissions across the globe are leading to climate change. The intensity of disasters in different forms is compounding the life and living patterns. In linking these concerns, the place of the poor and disadvantaged, particularly the women, needs specific stress, since women are economically the most vulnerable in society and very often the hardest hit by the harmful effects of environmental degradation, which destroys their assets, resources and livelihoods. From the foregoing, it is clear that the linkages between development and disasters are very complex.

Rationale

According to Mahmoud Mohieldin (2012),[2] Managing Director, World Bank, even though people are exposed to disasters, there need not be huge losses provided there is enough scope to focus on resources like information, incentives, infrastructure and good governance. Adequate and timely investments on information and

2. *Times India*, 14 March.

prevention should be made, and societies should be incentivised to take these concerns earnestly. There should be sufficient attention on the infrastructure required to avoid damages. Besides, good governance practices should be adopted to make things happen. The will to prevent, foresight to comprehend and ability to mitigate should come entirely from sincere governments as also the leaders' desire to pay attention to disaster mitigation.

Well-organised establishments alongside capable prevention measures should go hand-in-hand. To make things explicit, appropriate information subsuming early warnings on the type and place of disaster risks threatening to occur, incentives for encouraging investments in prevention and a dependable and accessible infrastructure can help ensure that natural hazards do not result in loss of lives and property. India needs to address counter-disaster measures, but while doing so, it does not need to panic with regard to incurring huge expenditure, as there are ingenious ways to ward off the perils. Simple measures involving well-timed information and prevention make sense. For illustration, greater losses were prevented in Japan by ensuring that every child is trained in earthquake drills.

Role of Civil Society Organisations (CSOs) in Disaster Management

The High Powered Committee on Disaster Management Plans (2001), recommended a nationwide networking of NGOs (inter-changeably used here as CSOs), for synchronised action encompassing vital facets of disaster management. However, the term civil society usually refers to a collection of institutions, where people meet, discuss and engage with public affairs, which of late includes disaster management. Put tersely, civil society subsumes NGOs, CBOs (community-based organisations), labour unions, international organisations such as Oxfam and even the welfare associations of local residents.

The Disaster Management Act (2005) gave Constitutional buttress by entrusting the State Executive Committee and District Authority to play a key role in disaster mitigation alongside voluntary action available at the cutting-edge level. The concept of civil society encompasses a wide range of organisations which in a broad sense may include all market and non-state organisations and structures in which people organise to pursue shared objectives and ideals. These include NGOs of all hues and shades (farmers, professionals, etc.), CBOs, formal and informal groups. Accordingly, a major shift in the role and responses of CSOs has occurred from its traditional strategy of supporting development issues to more specialised field like offer assistance in rescue, relief, emergency health care, i.e., the first aid, sanitation, hygiene, counselling and trauma care besides damage assessment, participative training and sensitisation campaigns as also in constructing temporary shelters for rehabilitation of disaster victims and bringing timely relief to the affected communities.

Furthermore, states with all kinds of resources at their disposal have the primary responsibility to respond to disasters effectively. But sheer magnitude and complexity of mega disasters like Bhuj Earthquake, Tsunami, Kosi floods (Bihar) and Krishna floods (Andhra Pradesh [AP]), to name a few recent major disasters, make it extremely difficult to respond effectively, particularly in the immediate aftermath of jumbo debacles. Generous help from outside agencies having requisite permission, expertise, experience and resources may lend a vital service in the state's efforts in managing major events more efficiently.

The CSOs in the pre-disaster phase must work for awareness generation, education, training, formation of village-level task forces/committees/teams, development of DM (disaster management) plans, carrying out mock drills and simulation exercises, vulnerability assessment, coordination between the government and non-government agencies. During post-disaster phase, the CSOs can take lead by providing technical and material support for safe re-construction, timely resumption of schools and restoration of livelihoods and help the government in swiftly monitoring implementation of diverse reconstruction and recovery programmes. Hence, it is imperative to encourage voluntary bodies working enthusiastically at the grass-roots. The larger CSOs/NGOs can provide technical and material support to the smaller ones, which have been working with the local communities and have better appreciation of their vulnerabilities, needs and capacity for revival to bring back to pre-disaster stage the soonest. A well-orchestrated approach giving a pivotal role to local CSOs/NGOs, has crucial relationships with the larger ones and the district administration augur well for institutionalising the role of CSOs in disaster mitigation.

PRIs and CSOs in Disaster Reduction

As a general rule, PRIs (panchayati raj institutions) at the grass-roots can play an important role in the process of transforming the community into a self-dependent entity while dealing with disasters. The CSOs (NGOs and CBO) have been working actively at the local level, i.e., at the GP (*gram panchayat*) and progressively at the next two higher tiers, i.e., *panchayati samiti* and *zilla parishad* as well.

The NGOs play a critical role in raising awareness among individuals and organisations involved in EWS (early warning system) and in its implementation, particularly at the community level. In addition, NGOs play an important advocacy role to help ensure that EWS to continue on the agenda of policy makers. On the other hand, the CBOs are created at the local level to meet the demand for service of a particular exertion as they establish a good rapport with the community and become instrumental while dealing with in all the three disaster phases.

The GP being the lowest rung of the formal rural local government system ought to provide an appropriate foundation for CSOs. In fact, local governments should have considerable knowledge of the hazards to which their communities are exposed. They must be actively involved in the design and maintenance of early warning systems and comprehend information received to be able to advise, instruct or engage the local population in a manner that increases their safety and reduces the potential loss of resources on which the community depends.

Despite the Constitutional buttress, PRIs have not been given an appropriate role in DRR due to lack of political will and bureaucratic indifference apart from absence of futuristic vision.

That said, elected *panchayat* leaders had striven manfully to help their villagers, who suffered during disasters, with whatever meagre resources they have had at their disposal.

Disaster Mitigation in Coastal AP: A Success Story

The Coastal Area Disaster Mitigation Efforts (CADME) (Meda n.d.), a network of 20 Voluntary Organisations in AP, initiated a programme titled: Community Induced Early Warnings and Preparedness, in over 350 most vulnerable villages in the east coast and established task forces at each vulnerable village to mitigate the effects of disaster as well as to help their brethren in the pre-disaster, during and post-disaster phases. Over 5,000 task force members were well-appointed with skills and understanding of disaster preparedness and are ready to render services anywhere in the country. More than 350 village level contingency plans were developed with the involvement of local communities and task forces were formed in all the adopted villages. Almost all the contingency plans identified vulnerable areas, communities at risk, evacuation plans, nearby elevated grounds as safer havens apart from earmarking the activities as well as responsibilities of task force members during different but interrelated disaster phases.

The CADME network has been using both the electronic (short films and video shows) and print media (posters and leaflets) in the schools to reach out to children of 50 villages in the East Godavari District of AP to help protect not merely themselves, but also others during disasters. This unique programme was based on the belief that children will be able to absorb and easily disseminate information to their parents and other members of the community leading to changes in attitudes of village elders. Thus, CADME has successfully involved local communities in developing their own village contingency plans to face adversity and in mitigating unfavourable effects of calamities. Currently, the community is well on track to encounter the problems cropping up during disasters.

What We Need to Know

Undoubtedly, the United Nations Development Programme (UNDP) (*The Hindu*, 24 May 2007) made a huge contribution for capacity building of nearly all stakeholders: i.e., the people; community; civil society and local government institutions on risk reduction practices. Its impact was evident during the post-Tsunami reconstruction, across all the coastal districts of Tamil Nadu. In the awareness generation programmes organised with UNDP's support, the villagers were educated on various forms of disaster risks, besides being oriented on how to be equipped to handle disasters aided by EWS, satellite/community radios, mock drills and simulation exercises.

As a follow-up to the UNDP's efforts, having a positive impact on the DM over the past decade in India, the World Bank funded a $410 million for the Emergency Tsunami Reconstruction Project (ETRP) in Tamil Nadu. The design focused to reduce the vulnerability of coastal communities through active involvement of CSOs, while constructing permanent shelters, multi-hazard early warning systems and livelihoods of people by way of CBDM (community based disaster management). In a similar vein, the World Bank has funded several other risk reduction projects like the National Cyclone Risk Mitigation Project (NCRMP) to a tune of $255 million, now being implemented in AP and Odisha by focusing on the last minute connectivity in the EWS and capacity development of government institutions and CSOs to ease out the administration diligence.

Along the way, Sastry (2002) maintains that such efforts inevitably entail additional time, resources on tap and satisfactory financial support on top. But the costs of some sporadic adversities and consequences of diverse mitigation measures would obviously outshine these efforts. Thus, it is obvious that the investments on community will have multiplier effects and high social cost-benefit ratio. Hence, it is not an exaggeration to assume that a rupee spent now on disaster preparedness will pay several hundred times in return in the long run.

Linking DRR with development is not about implementing a series of discrete disaster reduction schemes divorced from normal development plans, which ought to include mitigation measures in the overall developmental process. Besides these measures, one must not forget to integrate concerns like gender and environment into the development process. The Jawaharlal Nehru National Urban Renewal Mission (JNNURM) is a good example of including disaster mitigation alongside development measures. The DRR at the government level should go hand-in-hand with the practices of the community on the need to reduce the unacceptable risks from disasters in small steps. It must be ensured that people consider themselves as 'participants' rather than 'recipients' of the government sponsored development schemes.

Community preparedness and participation is one of the most important features of CDBM. There is little point in trying to prevent flooding which occurs regularly; instead, the system should become better equipped to manage it. The Kosi Floods (2008) in Bihar and the Krishna Floods (2009) in AP have exemplified how the women SHGs (self-help groups) managed community kitchens during relief activities. The community already has an indigenous in-built risk resilience system, which needs further strengthening. The *Guppedu Biyyam Pathakam* (a Telugu word meaning a fistful of rice contributed daily by each family) in the cyclone-prone East Godavari District of AP and the Karai system of grain storage using bamboo in the flood-prone Bihar epitomise such home-grown practices. Drought-prone areas in the country provide countless examples of local rainwater harvesting systems. Efforts should be made to use traditional resilience mechanisms, besides the external support that can be provided to the community for risk reduction.

Implications for Policy People

Stimulating inter-disciplinary and inter-sector partnerships and the expansion of DRR networking among governments at the national, state and grass-roots, greater involvement of officials and legislators, private sector, academic institutions and CSOs is of essence apropos DRR. This calls for strong coordination mechanisms like appropriate institutional structures for disaster preparedness, emergency response and early warning and the incorporation of disaster reduction concerns in the national and state planning processes. Also, efforts to link natural resource management with disaster reduction should be encouraged.

It has been widely recognised that DRR ought to become part of the total developmental perspective of an area. The National Disaster Management Policy (2009) precisely emphasised on this need for disaster preparedness and risk reduction to become an ingredient of the ToR (Terms of Reference) for any forthcoming project or developmental activity. The Mahatma Gandhi National Rural Employment Guarantee Act (MGNREGA) provides a precious opportunity for risk reduction in drought-prone areas in the country by executing works relating to soil-water-forest development and in the process strengthen the natural resource-base for sustainable development and poverty alleviation. The works taken up under this much applauded wage earning devise include construction of cyclone/flood shelters, modernisation of drainage system and land development and embracing gainful wage employment by executing poverty alleviation schemes.

According to Karnataka Minister for Panchayat Raj and Rural Development (Patil 2013), MGNREGA is one of the most important nation-building exercises that goes beyond the mandate of employment generation. Works taken up under this flagship programme in conjunction with other related government departments should be directed at making villages in Karnataka 'drought-proof.' He had also emphasised the need to harness the employment guarantee scheme to revive the sagging agricultural economy and make farming more productive. By the way, many schemes such as planting farm bund trees and reviving farm tanks would serve the purpose.

Critical Observations

Realisation of lofty ideals of MGNREGA, however, is hamstrung by rigmarole concerning the bureaucratic and political apparatchiki. The developmental needs for different disaster-prone zones should be assessed and specific planning approaches should be evolved, keeping in view the physical characteristics so that DRR and the socio-economic development of an area can be suitably integrated.

As for capacity development of the community in terms of knowledge, skills and attitudes, several models exist in India. There is a dire need to look at the strengths (uniqueness and social acceptability) and weaknesses (incompatibilities with the socio-cultural milieu and institutional arrangement and its abilities) of these models and to evolve an appropriate strategy instead of routinely training the stakeholders.

In the post-IT (information technology) epoch, the stakeholders need to be oriented to using the IT-enabled services (ITES) for speedier communication and effective implementation of DRR. Development of partnerships in both planning and execution of risk reduction measures would facilitate resource pooling and allocation of tasks juxtaposing the capability of the stakeholders' needs to be harnessed. In view of frequent occurrences of disasters in the developing nations including India, it is imperative to have a professional and an institutional approach with community orientation to effectively implement DRR and prevent/reduce risks and their incidence. It therefore follows that one has to move away from the dependency models toward the community-centric resilience ones.

Alternative Propositions

For the purpose on hand, three fresh propositions are presented concerning the role of CSOs in DM for consideration and refinement.

State Acting as a Key Player and Involving Others Like Civil Society Organistions, when Required

The model proposed here is the most common one that is being followed in the developing world. Here, the state will be taking lead in the planning and execution of the formal delivery system, consisting of officials and elected representatives responsible for pre-disaster preparedness and post-disaster relief and reconstruction. Perpetual dependency of the communities is one of the outcomes of such a model, which is the usual course followed in India too. Involvement of CSOs is not uncommon and keeping the strengths of these organisations, tasks will be assigned under continuous guidance and control from the above. It is obvious in such a model, transparency and accountability are minimal and the media is the major source of information on various activities undertaken by the government. Experience has shown that this model has been found to be more effective during the relief phase rather than in rehabilitation period. The ITES have also contributed to the centralisation of the entire decision-making process and resource control. Also, such models pay less attention to preventive measures since the delivery system is burdened with multiple tasks and decision-making is one among diverse responsibilities to concentrate on.

State Involving Stakeholders through Delegation of Tasks with Commensurate Powers

This model is observed in areas where there is frequent occurrence of disasters. Institutions with proven capabilities have been roped-in to performing certain tasks and entrusted with adequate powers and resources. In a way, such a model could be construed as a mid-path with partial institutionalisation of functions. However, the onus of planning and execution centres on the formal delivery system. The *modus operandi* of this model suffers from a major deficiency that could actually jeopardise the reputation of CSOs, when the quality of delivery is poor. Evidently under this model, disaster preparedness is institutionalised to some extent. The role of the community is largely passive and confined to relief seeking. On the other hand, one of the strengths of this model is that it enables the government to utilise competencies of the private sector and reduce to some extent the dependency on the formal delivery system, which is under the aegis of the state.

Stakeholders Taking Prime Role in the Entire Process but Seeking State Support

This model is consistent with the spirit of the Constitution 73rd and 74th Amendments aimed at strengthening PRIs and ULBs (urban local bodies), by means of a shift from delegation of powers to devolution of functions, functionaries and funds to the local self-governance institutions. Conversely, it is necessary that all stakeholders including national, state and local authorities to capitalise and have the will to make things happen. As such, evolution should be based on local capabilities and culture and take advantage of relevant funding, which remains a big challenge for the moment. The change envisaged through elected leaders of local governments provides for a handful of opportunities for the state to invest in community development through enhancement of capacities of human capital. However, planning and execution is largely decentralised with the community emerging as both the primary stakeholder and controller over resources and decision-making.

Accordingly, the community's vigilance and monitoring committees would also facilitate better quality of execution even while bringing greater transparency and accountability in the entire process. This model emphasises on both horizontal and vertical integration. The latter is one of the necessary conditions for the formalisation of partnerships among the formal and informal institutions to infuse a sense of ownership amongst the stakeholders. Under this decentralised model, the community alongside PPPs (public-private partnerships) will receive priority, and hence would strive to empower the communities, which is both a necessary and sufficient condition for disaster mitigation and sustainable development.

Furthermore, this model would help address the concerns of gender and the marginalised groups. Public purpose and private entrepreneurial dynamism need to be enmeshed to lend a helping hand to cope with catastrophe, besides ensuring competent management, efficient and effective operations *vis-à-vis* measurable results. Consequently, there is an evocative challenge for the corporate sector and the government to join forces. One could even visualise that this outline gradually lays focus on the preventive measures in order to lessen the unacceptable disaster risks and damages. As a result, it is expected that the ITES would help the stakeholders in terms of providing information on the last minute connectivity in early warning systems, community development, etc., and at the same time coping with disaster preparedness and reduction.

Conclusion

In fine, it may perhaps be apposite to infer that people in India, like in any other developing nation, comprise the largest and most readily accessible workforce. Due to the vital role people perform apropos several different kinds of occupations and avocations that the rural communities are engaged-in like the farming (including agricultural production, animal husbandry, etc.) and non-farming (construction of infrastructure like road and land development) sectors, while providing various forms of communications, people have been evenly adept in disaster mitigation too. By and large, this could be pursued either through the formal (local governments: PRIs and ULBs) or informal (non-government) groups including CSOs, the corporate sector or the society itself. Besides, people should be made aware of and facilitated to appreciate their responsibilities toward their own well-being. This could be accomplished by way of integrating dynamic factors in the developmental sphere with the community preparedness by taking advantage of the willingly available colossal resource base. Closing this piece on a hopeful note, we might suggest similar studies with a base of hypotheses progressively refined as we go along.

References

Guha-Sapir, D., F. Vos, R. Below with S. Ponserre (2010). *Annual Disaster Statistical Review*.

Meda, G.P. (n. d.), "CSOs and Coastal Area Disaster Mitigation: A Success Story", Paper presented at a *National Conference on Disaster Management*.

Patil, H.K. (2013). "Utilise MNREGA to Make Villages Drought-proof in Karnataka", *The Hindu*, 22 December.

Sastry, K.R. (2002). "Community Participation and People's Awareness in Disaster Management," *ASTI Journal of Training and Development*, July-December.

14

M. GOPINATH REDDY, B. SURESH REDDY
and P. STEVEN RAJ

Reduction in GHG Emissions through Sustainable Production of Biofuels: Is it a Viable Option?

A Review of Experiences

Introduction

Depletion of fossil fuels at an alarming rate has attracted increasing attention to blending biofuels worldwide. A debate between biofuels supporters and its opponents is intensifying. Given the rapidly increasing demand for energy which is projected to double by mid-21st century, it is expected that biofuels will become an important part of the global energy mix and make a significant contribution to meeting energy demand. Drivers for a wide introduction of biofuels vary across the world and include a broad range of issues from land-use to energy security, to economics and environment. India's energy demand is expected to grow at an annual rate of 4-5 times over the next couple of decades. According to the International Energy Agency (IEA), India will become the largest single source of global oil demand growth after 2020. Most of the energy requirements are currently satisfied by fossil fuels: coal, petroleum based products and natural gas. As domestic production can only bridge the gap by 25-30 per cent, this has serious effects on the energy security of the country added to the burgeoning burden of imports. India had indigenous oil reserves of 758 million tonnes in 2012-13 and its production was 37.86 million metric tonnes. During 2012-13, the total refinery throughput requirement was 219 million tonnes and the difference of 181.14 met through imports. In 2012-13, the country imported 185.0 million tonnes of crude oil (See Figure 14.1) which amounts to nearly 80 per cent of its domestic crude oil consumption and accounting to nearly 30 per cent of the country's total imports.

Figure 14.1

Crude Oil Imports in India Between 2004 and 2013

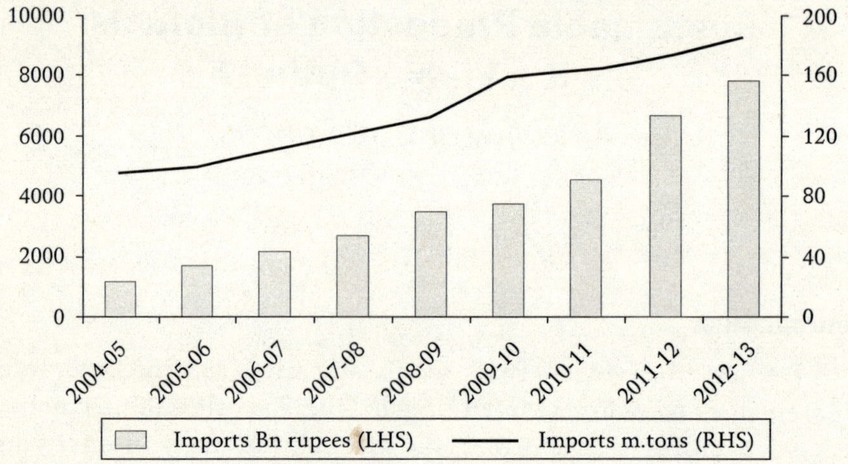

The rate at which the energy needs are growing demands either a greater reliance on imports (which is strain on depleting fiscal resources and foreign exchange) or a shift to alternative sources. This necessitates the change of focus towards biofuels as a favourable alternative option. In addition to providing energy security and a decreased dependence on oil imports, biofuels offer significant benefits such as reduced emission of pollutants and GHGs (greenhouse gases). Most importantly, the industry has a potential to create avenues to raise farmer incomes and restore degraded lands, while at the same time contributing to climate change mitigation.

India's primary energy use is projected to expand massively to deliver a sustained GDP (gross domestic product) growth rate of 9 per cent through 2031-32 even after allowing for substantial reduction in energy intensity. In order to fuel this on sustained basis, the growth of around 5.8 per cent per year in primary energy supply including gathered non-commercial energy such as wood and dung of would be required. Commercial energy supply would need to grow at about 6.8 per cent per annum as it will replace non-commercial energy, but this too involves a reduction of around 20 per cent in energy use per unit of GDP over a period of 10 years. India is confronted with energy crisis due to the depletion of resources and increased environmental problems. Diesel is the primary transport fuel of the country and comprises around 42 per cent of the total fuel market, majority of which comes through import market. With self-sufficiency levels in crude oil being a distant dream, there is a growing interest/need in development and commercialisation of a bouquet of alternative fuels. Keeping this in view, the Union cabinet of the Government of India (GoI) approved national policy on biofuels on 24 December 2009, which stresses on mainstreaming of biofuels in India to meet its ever increasing energy requirements and to limit the carbon footprint of the country. The policy calls for setting up of National Biofuel Coordination Committee (NBCC) headed by the prime minister 'to provide overall coordination, effective end-to-end implementation and monitoring of biofuel programmes." Another biofuel steering committee would be set up to tend to more regular and day-to-day of the same which would be chaired by cabinet secretary (GoI 2009). National biofuel policy aims at ensuring that the next generation of technologies is based on non-food feed stocks so as to avoid conflicts with food security. The policy aims at mainstreaming of biofuels and therefore, envisions a central role for it in the energy and transportation sectors of the country in coming decades. The policy aims at bringing about accelerated development and promotion of the cultivation, production and use of biofuels to increasingly substitute petrol and diesel for transport and can be used in stationary and other applications, while contributing to energy security, climate change mitigation, apart from creating new employment opportunities and leading to environmentally sustainable development. The policy sets an indicative target of 20 per cent blending of biofuels, both for biodiesel and bioethnol, by 2017. Presently, the ethanol blending with gasoline was 2.9 per cent, as off 2013. The GoI's ambitious plan of producing sufficient biodiesel by 2011-12 (marketing year October-September) to meet its mandate of 20 per cent blending with diesel was unachievable due to unavailability of sufficient feedstock (Jatropha seeds) and lack of high yielding drought tolerant Jatropha cultivars (Global Agricultural Information Network [GAIN] 2013).

In this context, the objective of this paper is to understand the viability of the large-scale biofuel cultivation and its role in reduction of GHG emissions. This paper consists of seven sections including this introduction. The second section two deals with the advantages and disadvantages of biofuels. The cultivation of biofuels and national mission on biodiesel are discussed in the third section. The fourth section describes the impact of biofuels cultivation on land use, environment and livelihoods of the communities. The search for alternative biofuels is presented in the fifth section. The debate of energy security *versus* food and fodder security is discussed in the sixth section and concluding observations are made in the final section.

Biofuels

Advantages

Added to its uniqueness as an environmental friendly fuel compared to either gasoline or petroleum diesel, biofuel is also recognised due to its portability, ready availability, renewability, higher combustion efficiency, lower sulphur and aromatic content, and higher biodegradability (Ma and Hanna 1999, Konthe et al. 2006). In addition, biodiesel can be used in any diesel engine without any modification. Blends up to 20 per cent biodiesel mixed with petroleum diesel fuels can be used in nearly all diesel equipment and are compatible with most storage and distribution equipment.

The clamour for shift to biofuel driven energy, especially in the transportation sector, is of late gathering ground more in the developing countries like India, given its potential to reduce the dependency on imported fuel and thus reducing the burden on the exchequer. More-over given that biodiesel can be manufactured from domestically cultivated crops would also contribute to better farm level incomes and also increased employment generation both at the field and factory level.

Table 14.1

Technical Properties of Biodiesel

Common Name	*Biodiesel*
Common chemical name	Fatty acid (m)ethyl ester
Chemical formula range	C14-C24 methyl esters or C15-25H28-48O2
Kinematic viscosity range (mm^2/s, at 313 K)	3.3-5.2
Density range (kg/m^2, at 288 K)	860-894
Boiling point range (K)	>475
Flash point range (K)	420-450
Distillation range (K)	470-600
Vapour pressure (mm Hg, at 295 K)	<5
Solubility in water	Insoluble in water
Physical appearance	Light to dark yellow, clear liquid
Odor	Light musty/soapy odour
Biodegradability	More biodegradable than petroleum diesel
Reactivity	Stable, but avoid strong oxidising agents

Source: Demirbas (2009).

Disadvantages

The major disadvantage of biodiesel are its higher viscosity, lower energy content, higher cloud point and pour point, higher nitrogen oxide (NOx) emissions, lower engine speed and power, injector coking, engine compatibility, and high price. The specific fuel consumption values of biodiesel are greater than those of commercial diesel fuel, while the effective efficiency and effective pressure values of commercial diesel fuels are greater than those of biodiesel. Despite their appeal as an alternative to fossil fuels, biofuels are also subject of considerable controversy.

The primary concern is that the growth of agricultural crops to produce biofuels may be inherently unsustainable as crops require land and water to grow and this would inadvertently in the long run result in the shift from food to non-food/fuel crops given higher incentives. Crops of any nature in industrialised agriculture require synthetic inputs such as fertilisers and pesticides, both of which are produced and transported using fossil fuel energy. This fact adds to the overall energy required to produce crops that provide energy and raises questions about whether the finished product provides more energy than is spent to produce it (Giampietro et al. 1997). There is also considerable debate questioning whether the end fuel product will truly be better for the environment than fossil fuels when subjected to a LCA (life cycle analysis) (Heintzman and Solomon 2009, Puppán 2002). LCA is defined by the International Standards Organisation (ISO) as "a compilation and evaluation of inputs, outputs and potential environmental impacts of a products system throughout its life cycle" (cited in Guinée et al. 2001).

Table 14.2

Biodiesel Emissions Compared to Conventional Diesel

Emissions	*B100 (100% biodiesel)*	*B20 (20% biodiesel)*
Regulated emissions		
Total unburned hydrocarbons	-93 per cent	-30 per cent
Carbon monoxide	-50 per cent	-20 per cent
Particulate matter	-30 per cent	-22 per cent
NOx	13 per cent	2 per cent
Non-regulated emissions		
Polycyclic aromatic hydrocarbons (PAH)	-80 per cent	-13 per cent
NPAH (Nitrated PAH)	-90 per cent	-50 per cent
Life cycle emissions		
Carbon dioxide (LCA)	-80 per cent	
Sulphur dioxide (LCA)	-100 per cent	

Source: GoI (2003).

Experiences worldwide suggest that the conventional fuels can be successfully substituted with biofuels. There are many successful experiences from all over the world from Canada in North America; Brazil, Argentina and Columbia in South America; France, Germany and the European Union; India, China, Indonesia, Malaysia and Thailand in Asia; and Australia. Over the last decade, between 2000 and 2009, boil fuel production has increased dramatically from 16.9

to 72.0 billion litres, while biodiesel grew from 0.8 to 14.7 billion litres. Of all the biofuel experiences, Brazil, using sugarcane-based ethanol, is been regarded as the most successful one as all gasoline sold in Brazil is a blend of 18-25 per cent ethanol. The national ethanol programme, Proalcool, was launched in 1975, after the second oil crisis in 1979, wherein Brazil decided to shift to cars powered by entire hydrous ethanol—this was very successful as by 1985, 95 per cent of the light vehicles produced in Brazil were built to use hydrous ethanol. In 2003, flex flues vehicles were launched and currently account for 90 per cent of the new sales, constituting the high point of Brazilian ethanol success story in the present decade. Brazil's ethanol programme was more consolidated given the fact that: i) gasoline contains 25 per cent of ethanol, ii) ethanol is available in all gas stations, and iii) 50 per cent of the car fleet and 90 per cent of new car sales are of flex fuel. Brazil produced 717 million tonnes of sugarcane, which yielded 36.1 million tonnes of sugar and 27 billion litres of ethanol. Most of the ethanol produced is absorbed in the domestic market, where it is sold as either ethanol fuel or blended with gasoline.

Biofuels in India

The two prominent biofuels in India are bioethnol (or simply ethanol) and biodiesel, which made from biomass containing sugar like molasses and vegetable oil like non-edible Jatropha oil respectively. The policy document on biofuels defines biomass as "biodegradable fraction of products, wastes and residues from agriculture, forestry and related industries as well as the biodegradable fraction of industrial and municipal wastes" (GoI 2009).

Ethanol is manufactured in India by fermentation of molasses, which is by-product of sugar industry. India is fourth largest producer of ethanol in the world, after Brazil, the US and China, with distillation capacity of 2,900 million litres per year. GoI made 5 per cent blending of ethanol with petrol mandatory in nine sugarcane producing states in September 2002. However, due to supply shortage, the mandate was made optional in October 2006. In October 2007, the government again made it mandatory to have

5 per cent ethanol blend in petrol across the country with exception of Jammu & Kashmir (J&K), the Northeast and island territories. Now, the policy on biofuels has an ambitious target of 20 per cent blending by 2017 (see Table 14.3).

Table 14.3

Projected Demand for Petrol and Diesel and the Biofuel Requirements of India

Year	*Petrol Demand in Mt*	*Ethanol Blending Requirement (in MT)*			*Diesel Demand in Mt*	*Biodiesel Blending Requirement (in MT)*		
		@ 5 (%)	*@ 10 (%)*	*@ 20 (%)*		*@ 5 (%)*	*@ 10 (%)*	*@ 20 (%)*
2006-07	10.07	0.50	1.01	2.01	52.32	2.62	5.23	10.46
2011-12	12.85	0.64	1.29	2.57	66.91	3.35	6.69	13.38
2016-17	16.40	0.82	1.64	3.28	83.58	4.18	8.36	16.72

Source: Planning Commission (2003). *Report of the Committee on Development of Biofuel.* Government of India. 16 April.

The biodiesel industry, however, is not as mature and is still in incubation stage. The demand for diesel is four times the demand of petrol in India. Considering this and other cost associated with conventional diesel fuel, GoI formulated the National Biodiesel Mission in 2003.

National Biodiesel Mission

National Biodiesel Mission was proposed in a Planning Commission report of the Committee on Development of Biofuel, with an aim of meeting 20 per cent of fuel requirements of the country with biodiesel by 2011-12. After extensive research Jatropha seed was considered feasible for oil extraction in Indian Biodiesel Mission. The mission was to be implemented in two phases: Phase I (2003-2007) would be consisting of demonstrating Project that would lay foundation for self-sustaining and fast-growing biodiesel programme. It would concentrate on producing enough feedstock for production, testing the viability of processes, and to inform and educate the potential participants. Phase II of the mission would produce enough vegetable oil from feedstock to sustain the 20 per cent mandate of biofuel blend by 2012 (GoI 2003). Large-scale cultivation of Jatropha must be established before biodiesel production can meet even a 5 per cent blending requirement nationally. But, amid reports of unviability of Jatropha seed and overall negative energy balance of biofuel processes, the National Biofuel Mission and policy recommendations seem to hang in jeopardy (Negi et al. 2006, Gonsalves 2006, Singhal and Sengupta, 2012).

Biofuel Cultivation: Impact on Land Use, Environment and Livelihoods

According to the biofuel committees' norms of per hectare yield of Jatropha oil seeds, the biodiesel yield of Jatropha oilseeds and the specific gravity of biodiesel, the total land requirement works out to 31.7 million ha. In reality, this kind of wasteland availability is possible in the field or not is a question of debate. The government describes wasteland as degraded land which can be brought under vegetative cover with reasonable effort, and which is currently underutilised and/or land which is deteriorating for lack of appropriate water and soil management or an account of natural resources. There are lands which are less productive, but there is no land which is waste in the sense that it easily lends itself for cultivation of this magnitude (Negi et al. 2006). Secondary literature review reveals that in order to achieve the high yields and reduce the cost of production, better land is being sought. The study done by Negi et al. (2006) found that one of the land which is being targeted for Jatropha is pastureland. With the better quality of land being targeted, there is bound to be an impact in terms of fodder reduction. The Rajasthan Government's move to privatise the commons was resisted and the amount of land to be allotted was scaled down from 64 lakh to 1 lakh with the safe guard that the community approves of such allotment. Singhal and Sengupta (2012) reported that Jatropha plantations will involve higher paid-out expenses to be incurred by Indian farmers, thus restricting potential benefits of the development of an agro-based energy producing industry to only a section of the farming community having access to credit at reasonable cost. They concluded that a major thrust on the development of biofuels as alternative fuels for Indian transport sector can pose a challenge to the security of food and non-food agricultural supplies.

A major debate continues world over about biofuels production and its impact on traditional agriculture, i.e., the perceived competition for land and the risk of displacing production of human and animal food by biofuels. Although land devoted to fuel production could reduce land available for food production, this is at present not a serious problem. In the long term, lignocellulosics are likely to become the primary source of biofuels. It is important in each particular case to evaluate the sustainability of raw material production to ensure that biofuels are developed in areas that do not affect the use of the basic resources of agricultural ecosystems such as soil, water, air and biodiversity (World Energy Council 2010). Although biofuels for aviation and shipping seem the most likely solution, the implications for land use are enormous for the development of road transport biofuels (Philp et al. 2013).

Need for Alternative

Feasibility of biofuels substituting fossil energy is possible only if the large-scale production of biofuel is biophysically feasible (i.e., not constrained by the availability of land and fresh water sources for energy crop production); environmentally sound (i.e., does not cause significant soil degradation, air and water pollution or biodiversity loss); and compatible with the socio-economic structure of society (i.e., requires labour productivity that is consistent with the existing labour supply and per capita energy consumption in society). The latter two conditions imply that biofuel system must deliver a sufficiently large amount of net energy to society per hour of labour employed in the cycle of biofuel production to make the process economically convenient for society while generating a sufficient low environment loading per net energy supplied to keep the process environmentally sound.

Efforts to bring in Jatropha as a viable alternative have failed as it has been difficult to initiate large-scale cultivation given the fact that it is not economically viable enough (Gonsalves 2006). An alternative fuel to conventional fuels must be technically feasible, economically viable, environmentally acceptable and easily available. Indian land use is primarily dominated by agriculture and hence it generates a large quantity of agro residues.

Given the limitations with regard to Phase I and Phase II crops, the Indian biofuel policy and an alternative approach are the need of the hour which would focus on multipurpose, short-duration annual crops that can either simultaneously yield fuel along with food and/or fodder or can be cultivated in rotation with food crops. It would enable even the small private farmers to derive benefits from the opportunities, while at the same time making it economically viable due to the large-scale cultivation given the alternative crops greater reach pertaining to increase in acreage. What is the need of the hour is practicable type of cultivation that is compatible with the routines and the possibilities of local farmers, rather than high-tech agrarian methodology that needs maximum inputs to deliver bumper yields. In addition to sugarcane, several alternative feed stocks such as sweet sorghum and tropical sugar beet are being tried along with lignocellulosic materials like wood and crop residues. In recent times, dry land staple food and fodder crops like sorghum and *bajra* are being explored by few institutions.

Sorghum and millets occupy a prime place in smallholder farming systems in arid and semi-arid regions, providing employment, income, food for consumption and feed for livestock. India has over 62.29 and 72.53 lakh ha of land under cultivation of Sorghum and pearl millet producing 52.56 and 81.49 lakh tonnes respectively as on 2012-13. Both these crops are cultivated as dual purpose crops; besides food grains, millet and sorghum are an important feed for livestock, especially in the dry months when other feed resources are in short supply (Nagaraju et al. 2013). The characteristics of these crops—rapid growth, tolerance to abiotic stress, greater water efficiency, limited fertiliser inputs, greater biomass and most importantly wide adoptability given their cultivation as a part and parcel of fodder economy—have made them an attractive feedstock for ethanol production. In addition, these are C4 crops having carbon fixing properties (that is, they are climate-change compliant). Given moisture stress, millets are the best alternatives for extreme weather conditions and are well suited to drought-prone regions (Rao et al. 2006). Ethanol producing companies, research institutions, governments and agricultural universities can coordinate with farmers to strategically develop value added utilisation of these crops to significantly increase energy security. Indian land use is primarily dominated by agriculture and hence it generates a large quantity of agro residues. These residues in the form of biomass could be utilised for energy production; sorghum and pearl millet would fit into this and could offer a remunerative crop to farmers as well as provide clean energy.

Figure 14.2

Area under Sorghum and Pearl Millet between 2001 and 2011

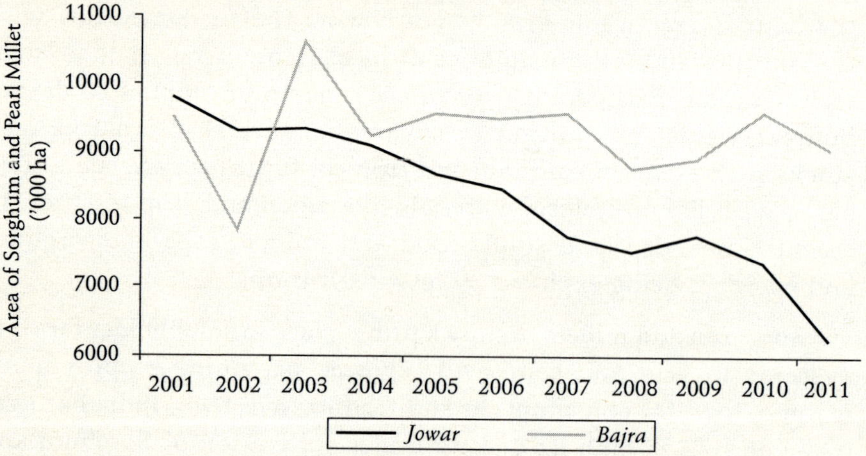

Economic Suitability of Sorghum and Pearl Millet

In India, cane molasses is the only major raw material for ethanol production. Molasses, which is a by-product of the sugar industry, contains 45-50 per cent fermentable sugars. Because of the ease of the process, this can be fermented into ethanol. However, there is growing interest for alternatives triggered due to acute water shortages all over the county where sugarcane is grown—the water requirement of sugarcane is about 36,000 m3 ha which creates huge water stress. A lengthy growing period and high water requirements are the disadvantages of sugarcane and here these crops can be a viable alternative.

Figure 14.3
Cost of Cultivation of Sorghum

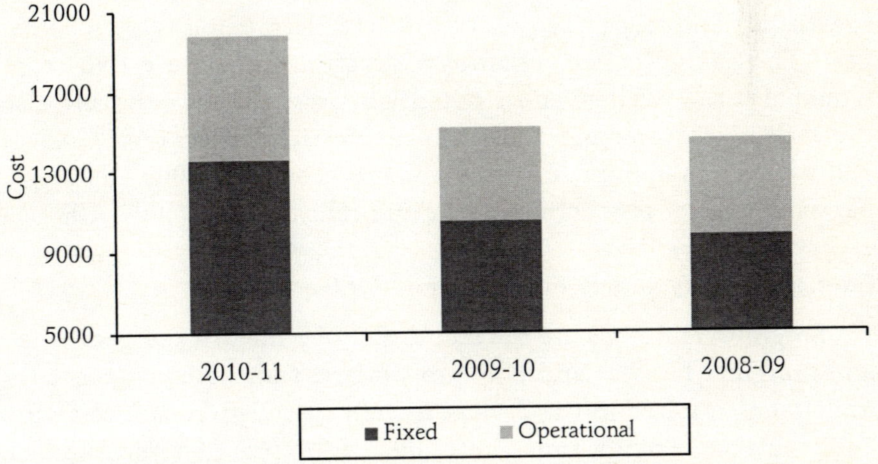

Figure 14.4
Cost of Cultivation of Pearl Millet

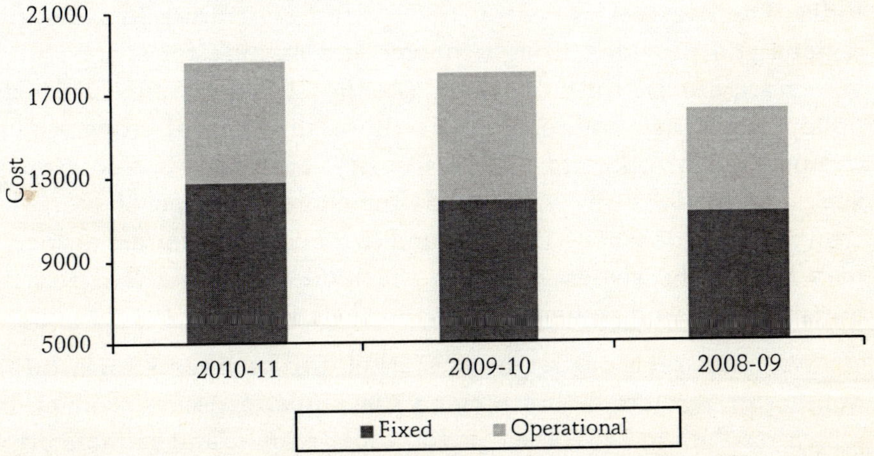

Sorghum and pearl millet are suitable for India's vast dry tracts with limited irrigation; it also requires minimum purchased inputs of ₹12,000 to ₹15,000, as against ₹41,750 to ₹48,250 per ha for sugarcane. The crops have a month cycle, permitting two crops per year. The energy cost of ethanol production from these two crops would also be lower, given its potentially lower input costs and also the lack of requirement for pre-fermentation processing. Cost of cultivation of sorghum and pearl millet is around ₹12,000-16,000 per ha, while cost of production of one quintal is around ₹800-1,100.

Energy Security versus Food and Fodder Security

Sorghum and *bajra* are the staple diet of a majority of poor small and marginal farmers and poor consumers. Poor policy support for millets has resulted in the decline in area of millet crops (See Figure 14.2). Longer cooking times, difficulty in preparation and the lack of value addition contributed to decline in the consumption of sorghum and millets. Extension efforts towards the cultivation of millets were relegated, and coupled with market failure—this has led to a failure to capture the nutritive value of millets (Nagaraju et al. 2013). In this context, the use of sorghum and *bajra* for biofuels would not only jeopardise the food security of millions of poor farmers, but also fodder security of millions of large ruminants in India. The livestock is central to dry land agriculture in semi-arid regions and provides drought power and manure for soil fertility which are crucial for crop yields (Reddy 2011). National biofuels policy emphasise not to promote those agricultural crops which compete with food and fodder security. Similarly, it also clearly spelt out to use only waste lands for the cultivation of biofuels. However, promotion of sorghum and *bajra* crops by certain agencies may have a disastrous effect on the food and fodder security of human beings and animals in India in the long run.

We have seen earlier how market forces have seen a large number of farmers falling prey to the unplanned over-production which resulted in lack of access to proper market and suitable price like in case of *safed moosli* growers and *doob* grass and also at times stranded by companies without any support by the government. The study done by Negi et al. (2006) in India highlighted the plight of Eucalyptus planters in many parts of the country who had planted the trees on their agricultural land on a buy-back guarantee from the paper companies in not a distant past when at the time of harvest, the companies imported pulp from Indonesia citing higher cost of purchasing the timber from the farmers as compared to importing the pulp. They felt that the lack of stringent regulations to enforce the contract terms and apathy from the government towards the farmers to protect their interests and efforts has resulted in severe losses on the part of farmers. Analysis of earlier interventions show that invariably one of the crop/species is promoted based on the projections as wonder crop for poverty alleviation citing reasons such as the potential of large-scale use in industry, good market price and usage of vast tracts of unproductive land, converting unusable tracts as valuable resources. Such interventions gain momentum from the projections focusing on the commercial aspects, citing the suitability of the system for large-scale commercial exploitation. Many times these projections supported by research findings are promoted by some agency. Largely, these projections do not take into consideration various environmental, economic and social aspects (Negi et al. 2006).

Conclusion

Achieving the energy security of the country by alternate ways is an important area being focused by the Indian policy makers. However any attempt to promote the use of major staple food crops such as *jowar* and *bajra* for biofuels production has long lasting impact on the food, fodder and nutritional security of millions of people and livestock in India. Large-scale biofuel production is not an alternative to the current use of oil and is not even an advisable option to cover a significant fraction of it (Giampietro et al. 1997).

References

Demirbas, A. (2009). "Progress and Recent Trends in Biodiesel Fuels", *Energy Conversion and Management* 50: 14-34. Elsevier.

GAIN (2013). *India: Biofuels Annual*. Global Agricultural Information Network .

Giampietro, M., S. Ulgiati and D. Pimentel (1997). "Feasibility of Large-scale Biofuel Production" *Bioscience* 47(9): 587-600.

GoI (2003). *Report of the Committee on Development of Biofuels.* New Delhi: Planning Commission.

———. (2009). *National Policy on Biofuels.* New Delhi: Ministry of New and Renewable Resources.

Gonsalves, B.J. (2006). *An Assessment of the Biofuels Industry in India.* United Nations Conference on Trade and Development/DITC/TED/2006/6.

Guinée, J.B., G. Huppes, A.W. Sleeswijk, M. Gorrée, R. Heijungs, R. Kleijn, A.M.M. Ansems, P.G. Eggels, R. van Duin, H.P. de Goede, H.A. Udo de Haes, R.M. Lankreijer, A. de Koning and Sangwon Suh (2001). *Life Cycle Assessment: An Operational Guide to the ISO Standards*. Maastricht, The Netherlands: Ministry of Housing, Spatial Planning and the Environment (VROM); Centre of Environmental Science, Leiden University (CML).

Heintzman, A. and E. Solomon (2009). *Food and Fuel: Solutions for the Future*. Toronto: Anansi.

Knothe, G., C.A Sharp and T.W. Ryan (2006). "Exhaust Emissions of Biodiesel, Petrodiesel, Neat Methyl Esters, and Alkanes in a New Technology Engine", *Energy Fuels* 20: 403-08.

Hanna, M.A. (1999). "Biodiesel Production: A Review", *Bioresour Technol* 70: 1-15.

Nagaraju, N., G. Basavaraju, R.P. Parthasarathy, C. Bantilan and S. Haldar (2013). "Sorghum and Pearl Millet Economy of India: Future Outlooks and Options", *Economic and Political Weekly* XLVIII(52). Mumbai.

Negi, K., J.K. Komal and P. Ranjan (2006). *Promotion of Biofuel in India: Issues and Prospects.* New Delhi: Society for Promotion of Wastelands Development (SPWD).

Parthasarathy Rao, P.P., P.S. Singhal, B.V.S. Reddy, K.N. Rai and S. Ramesh (2006): "Diagnostics of Sorghum and Pearl Millet Grains-based Nutrition in India", *International Sorghum and Millets Newsletter*.

Philp, J.C., K. Guy, and R.J. Ritchies (2013). "Biofuels Development and the Policy Regime: Biotechnology Policy Series", *Trends in Biotechnology* 31(1).

Puppán, D. (2002) "Environmental Evaluation of Biofuels", *Periodica Polytechnica Ser. Soc. Man. Sci.* 10(1): 95-116.

Reddy, B. Suresh (2011) "Dynamics of Soil Fertility Management Practices in Semi-Arid Regions: A Case Study of AP", *Economic and Political Weekly* XLVI(3): 55-63. Mumbai.

Singhal, R. and R. Sengupta (2012). "Energy Security and Biodiesel: Implications for Land Use and Food Security", *Economic and Political* Weekly XLVII(40). Mumbai.

World Energy Council (2010). *Biofuels: Policies, Standards and Technologies.* London

15

SHALINI SHARMA

Environmental Regulations and Sustainable Development

Introduction

At the Earth Summit in Rio de Janeiro in 1992, 178 governments signed the Rio Declaration affirming the principle that environmental decisions are best made with participation of all relevant stakeholders. Principle-10 of Rio Declaration lays the foundation for three pillars of good environmental governance: transparency, inclusiveness and accountability (Pring and Pring 2009).

India was operating on Environmental Protection Act (EPA) (1986) and Environmental Protection Rules (EPR) (1986). After Rio de Janerio 1992 Summit, Ministry of Environment & Forest (MoEF) and Government of India (GoI) had released Environment Impact Assessment (EIA) Notification 1994 for discharging the internationally agreed obligations under Rio Declaration. EIA Notification 1994 was the directive for providing EC (environmental clearance) to any new project or modification/expansion of old project.

Further due to requirement of development, changes in the size of projects, categorisation of projects on basis of the size of operations and making the procedure to flow in time-bound manner, EIA Notification 1994 was modified as EIA Notification 2006. These directives as laid by MoEF, India came into force on 14 September 2006. This provides the guidelines for obtaining the EC, that is mandatory for implementing any new projects/modification or expansion of old projects. This is to minimise the impacts of industrial project on environment and society, for fulfilment of legal requirements for operationalising the project and for project proponent to commit the effective environment management plans (as per suitability of industry/operations).

EIA Notification 2006 divided the projects into Category A, Category B1 and Category B2 depending on their size. Category A projects require EC by the MoEF, while Category B projects require clearance from State Environment Impact Assessment Authority (SEIAA) at State Level Pollution Control Board. Category B projects are divided in two categories i.e., B1 and B2. While B1 needs to undergo the process for obtaining EC, category B2 projects do not need EC for implementation of project. This leads to huge possibility of tailoring the facts to turn category B1 project to B2, therefore avoiding the screening at government level as well as the public hearing and stakeholder consultation.

National Voluntary Guidelines (NVGs) on Social, Environmental & Economic Guidelines released by the Ministry of Corporate Affairs (GoI 2009) recognised the dire need for business to focus on the 'Triple Bottom Line'—people, planet and profit to be socially relevant together with actualising their *raison d'etre* i.e., profit maximisation. The guidelines have been articulated in nine principles (NVGs, GoI, 2009) for the basic requirements for business to function responsibly, thereby ensuring the wholesome and inclusive process of economic growth in the country.

Guidelines on (i) Research and Development, and (ii) Sustainable Development (SD) for Central Public Sector Enterprises (CPSEs) released by the Ministry of Heavy Industries & Public Enterprises, GoI, (September 2011) has detailed about the natural resources and all environmental components that need to be conserved and used efficiently. It also provides the process of monitoring, target setting for SD and methodology of evaluation for implementing the SD practices in heavy industries and related sectors.

NVGs and Guidelines on SD are applicable to mining sector. Following study with example of mining sector provides the details about importance and ways for effective implementation of the above stated directives and guidelines.

Environmental Impact Assessment and Management Plan for Sustainable Development

Preparation and implementation of the Environmental Impact Assessment (EIA) and the Environment Management Plan (EMP) is the integral part of any new project or activities, expansion or modernisation of existing projects/activities listed in Schedules of EIA notification 2006. EIA Notification 2006 and related revisions provide directives for obtaining the EC for any project. Assessing the environmental impacts and preparing the management plans accordingly, however, has become business as the usual practice for obtaining the clearance. Any new project implementation in outskirts of city, villages or in rural areas may have wide potential impacts on lifestyle of local population, villagers, livelihood, biodiversity, ecology and environment of project site and surrounding areas for which EIA and EMP report is suppose to be prepared for obtaining EC and subsequently needs to be implemented effectively. If village population constitutes SCs (schedule castes) or STs (schedule tribes) then management plan for affected local people becomes even more difficult as many other social bodies/NGOs etc., get involved. If the land acquired for the project is agricultural land then EIA and EMP report should be strong enough to nullify the negative impacts on food and crop production. Controversial projects like Vedanta Mines (Saxena et al. 2010) and Taj Corridor in India caused displacement of population and degradation of food and forest production and therefore faced serious opposition during its implementation. The fact that the new projects are helpful in employment generation and economical development of society cannot be neglected. Serious flaw that exists in whole system is the non-implementation of EMP measures, as stated in the report submitted for obtaining EC.

In detailed study of EIA notification 2006 and working experience on real time EIA and EMP of open-cast iron ore mining projects during providing consultancy services (while working in B Ana Labs Ltd., 2006-07), the following loopholes are identified:

1. Effectiveness of the system for EC is weak.

2. Handing over of the responsibility of granting clearance of large number of projects to the state governments without any system of checks and counter checks is a loophole and needs to be looked into for status of discrimination for passing the project.

3. Exclusion of large projects (e.g., <500 MW thermal power plant, Solar power projects or cement production <1 MTPA) and construction (e.g., <20,000 sq m built-up area) projects from EIA clearance procedure. This will lead to more degradation of environment and depletion of natural resources. For example, land acquisitions for huge solar power plants may lead to unsustainable way of providing clearance.

4. Issues related to monitoring and compliance need to be addressed, specially because EIA Notification 2006 does not make regular monitoring mandatory. Also, on site assessment and post-project monitoring are entirely missing and not given in regulation. GoI has not created any monitoring agency with its clear role (e.g., SEIAA/self/private agency).

5. Project implementer is not bounded to provide complete information to the public residing/stakeholders around the project area. Project proponent may provide general draft EIA report to public before the public hearing. This leave lot of scope for diluting the project information and leading to environmental degradation, especially for dumping the waste material, etc.

6. Structure of Environmental Clearance Committee at State level, is not clearly defined, which may lead to inadequate expertise availability, resulting to clearance to ineligible projects.

7. EC, project tenure and tenure of EC provided to any project is not made available publicly/on website.

8. Grant of EC for 5-30 years and that too in absence of mandatory regular monitoring/post-project implementation

monitoring is a big lacuna leading to degradation of natural resources, other environmental and social components and increasing the chances for illegal activities.

9. Clear guidelines for impact assessment studies and planning effective management measures for ecology, biodiversity and indigenous community are completely missing.
10. No standards/guidelines are available to check the quality of EIA and EMP, therefore providing incomplete information for availing flexibility for obtaining EC is becoming a common practice.
11. Committee members involved with Ministry for approving EC remain open to risk.

NVGs (2009) and Guidelines for Sustainable Development for PSUs (public sector undertaking) (2011) also need to be integrated with EIA Notification 2006 and have to be followed by businesses and PSUs. This shall help in doing business in sustainable manner and ultimately in the sustainable development of country.

Professionalising the EIA and EMP Preparation and EC Procedure: Seven Tools for Implementation of Sustainable Business Projects

Study suggests seven methods/tools for effective implementation of EIA Notification 2006 in integration with NVGs (2009) and Guidelines for Sustainable Development for PSUs (2011).

Third Party Assessment of EIA and EMP Report of All Projects

Positive and negative aspects of any new project/activities/ project expansion should be assessed thoroughly. A third party (independent assessor) named as PV (project validator) should assess EIA and EMP report independently. PV should also conduct onsite visit and post-project monitoring. PV should have no involvement with the project implementer or State/Central Environmental Clearance Committee members. This independent third party should be employed on the basis of their qualification and experience and should work under the MoEF directly. Every EIA and EMP project should pass through the PV and they should be responsible for submitting the independent assessment (confidential) report to only Chairman of respective EC committee with a copy to highest authority of MoEF. The PV should be responsible for assessing the real time data report through onsite assessment.

MoEF should accredit the PVAs (project validating agencies). A process of screening and accrediting the agencies and experts shall need to be established. Internationally successful accreditation process is available for reference and the same can be followed, after making necessary required changes to suit national requirements, for accrediting the PVAs and PVs. The accreditation process may be referred in the "Handbook of Accreditation of Operational Entities by the Executive Board of CDM" (*http://cdm.unfccc.int/Reference/Guidclarif/accr_handbook.pdf*, United Nations Framework Convention on Climate Change [UNFCCC]).

PV should be given the right to assess the EIA and EMP project and site of project implementation. While assessment, PV should focus on requirement of project, biodiversity, population, economics, ecology and also the public opinion. PV should be competent enough to give their suggestions about the positive and negative impacts of upcoming project/project expansion. During annual monitoring and post-project monitoring, assessment of implementation of the environment management plan (as per project plan) and sustainable development should be the aim of PV.

Publishing the EIA and EMP Report and Validation Report on MoEF Website

Validation report should be treated confidential until the final decision about EC. Publishing the EIA and EMP report (including related documents) on MoEF website will be the appropriate methodology to make information available to public in transparent manner. After providing EC, EIA and EMP report, validation report and EC (detailing definite deadline of EC) should be published on MoEF website. This will help in professionalising the complete process of obtaining the EC, promoting the practices of ethical business and in achieving the sustainable development.

Public Involvement in Public Hearing

Public hearing, though mandatory for all new projects/ project expansion, is generally conducted as formality. "Laying down the rules for public hearing" can be the effective tool for implementation/operation of any project. Public hearing needs be given such importance because only local people know the best about the local ecology and economic conditions. The rules may include the provision of having at least five villagers from each village in 20 km radius of the project site. Their opinions should be recorded and authenticated. Out-of-context opinions may be given the proper justification for closing of issues.

Gram panchayat (GP) and *nagar panchayat* should be the part in decision making about the implementation of new projects. As per new Government schemes, GP (village head) is being given salary for implementing the government schemes, which means that *pradhan* (village head) will be instrumental in making decision resulting into positive or negative impact on ongoing/new projects.

Mandatory Post-Project Monitoring

There is no provision of post-project monitoring in EIA Notification 2006. India does not have stringent rules for implementation of this most important aspect of any industrial projects. Laying down the rules for 'post-project monitoring' that should be followed by project proponents is inescapable necessity. This will make the PP to follow the environment management of the project site and will help in making the PP responsible for post-project site rehabilitation/development. The PV nominated by MoEF should be assigned the job of annual onsite visit and post-project monitoring followed by submission of independent confidential report to MoEF.

Any major deviation in EMP plan, if noticed by PV, should be penalised. This shall ensure the implementation of management plan as stated during the EC.

Fast Green Bench

To solve the issues related to environment a 'Fast Green Bench' should be formulated. India is the third country in the world to start a National Green Tribunal (NGT), a judicial body exclusively meant to judge environmental cases. NGT initiated operations on 18 October 2010, but more than 6,000 cases are pending therein. Fast Green Bench can be structured to solve the smaller issues within 21 days of time period. Formulating the group of environmental experts comprising scientists, engineers, ex-industrialists and NGOs and empanelling them with the government for expediting the case in 'Fast Green Bench' to solve the case within 21 days, is the model suggested for successful and effective implementation of EIA Notification 2006. This panel will help the government by finishing off the EIA related case within 21 days. Environmental lawyers may be the empanelled lawyers for working on cases with Fast Green Bench. Existing judicial procedures which are inherently time consuming and are loaded in favour of delays (millions of cases are pending in courts of India) need to be done away with.

Standard for EIA and EMP Preparation

Standards like ISO 14001:2004 and OHSAS 18001:2008 exist, but it is not mandatory for companies/industries to follow them. Legal procedures like obtaining EC for any industrial projects/green field project/expansion should be standardised in order to restrict the natural resource depletion. Standard for preparing EIA and EM plan should be prepared and it should be made mandatory for companies to follow for obtaining the EC.

Online Committee Meeting for Environmental Clearance

MoEF introduced online submission of EIA and EMP reports. This need to be introduced at State level also. Step ahead, the process of monthly meeting of various committees for approving EC to any project involve huge risk—in case of disapproval of project, specially because of comments from any specific committee member. This is huge barrier for sustainable development. Security of committee members is the prime concern. Online monthly meetings for EC can solve this problem effectively. Committee members are renowned scientists/professors/busy professionals and online meetings can solve the challenges like time crises as well as enhance the security of members contributing fairly in the process of environmental clearance.

Results

Study on loopholes in above-stated polices and implementation of seven tools may help in professionalising the EC procedure and facilitating the implementation of sustainable development practices in businesses.

1. Professionalising the environmental clearance process by actualising the independent validation of projects seeking EC. PPP (public-private partnership) model for working as accredited validating agencies and validators can be implemented.
2. Formulating the concrete rules for public hearing facilitating the structured involvement of local level bodies like GP (village level body) and *nagar panchayat* (town level body) in the process.
3. Laying down the rules for annual project monitoring and post-project monitoring and penalising any deviation. This will make the project proponent to follow the environment management for the project site and will help in making the project owner responsible for post-project site development, which shall help in sustainable development.
4. An EIA and EMP standard (like ISO 14001:2004, etc.) should be prepared for providing guidelines for preparation of EIA and EMP and for assessing the quality of EIA and EMP report. This standard should be made mandatory to be followed by project proponent/EIA and EMP consultants. In this standard, each type of project shall be provided with certain guidelines for preparing the EIA report for taking mandatory environment friendly actions. Assessing the current situation of water, energy, health, agriculture and biodiversity should be made mandatory and should be assessed against this standard.

5. Third party assessment (project validation by independent organisation) should be made mandatory for all projects. Confidential and independent annual site study/project study/post-project monitoring report/EMP implementation report should be provided to MoEF and project proponent regularly for taking action, if any (once EIA and EMP standard is accepted nationally, then validation for project quality assessment can be done against the standard).
6. State level pollution control board need to be strengthened with enough competent manpower, who can do the works like lab testing, site visits, and should be made strong enough to file the legal case against project proponent in case of violation of rules. This may be on the basis of "the onus of proving innocence should be on accused" in case of violation of laws. Police or legal authority should be able to take charge of site immediately after filing of such cases.
7. 'Fast Green Bench' be created to give final decision within 21 days, on such cases.

Conclusion

Environmental regulation may need to be strengthened as per the requirement of time and increase in development pace of country. Practical work experience gained by preparing 21 EIA and EMP reports for obtaining environmental clearance from MoEF, GoI, was utilised to prepare this case study. Study suggests nine ways and means to achieve and implement the 'environmental regulation for sustainable development.' This policy research suggests innovative methods for making the project owners responsible for implementing the environment management plan effectively and contribute in sustainable development.

Government departments face challenges like lack of expertise/ manpower, etc. Private sector involvement with accountability, for sustainable development is, therefore, required. Successful PPP model and standards for EIA and EMP may be upscaled for practising sustainable development.

This paper detailed about:

1. Ways and means for professionalising the environmental clearance process by actualising the independent validation of projects seeking EC. Third party assessment shall be made mandatory for all projects.
2. PPP model for working as validating agencies and validators.
3. Laying down the rules for annual project monitoring and post-project monitoring and penalising any deviation
4. An EIA and EMP standard (like ISO 14001:2004) shall be prepared for providing guidelines for preparation of EIA and EMP reports and for assessing the quality of these reports.
5. Committee members involved with Ministry for assessment of projects for facilitating environmental clearance shall be able to provide fair outcomes.

Successful result can be upscaled in other countries also.

Acknowledgement: This study was conducted while working on real time EIA and EMP consultancy projects (for open cast iron ore mines) in Goa and Bellary District of Karnataka, India. No dedicated source of funding was available for conducting this policy research. Author thanks M/s B. Ana Labs Limited at Hyderabad, India, for providing the opportunity to work on iron ore mining projects. Thanks are also due to the then Director General, Engineering Staff College of India, Hyderabad, for supporting policy research.

References

B. Ana Labs Limited (2006-07). *Consultancy Projects for Preparation of EIA and EMP Reports for Open Cast Iron Ore Mines in Goa and Bellary Districts, India.* Unpublished 21 Reports: Real Time Projects Submitted for EC.

Pring, G. and C. Pring (2009). *Greening Justice: Creating and Grooming Environmental Courts and Tribunals: The Access Initiative.* World Resource Institute. p.x.

Saxena et al. (2010). *Report of Four Members Committee for Investigation into Proposal Submitted by Orissa Mining Company for Bauxite Mining in Niyamgiri,* Ministry of Environment & Forest, Government of India.

UNFCCC (2003). *Handbook of Accreditation of Operational Entities by the Executive Board of CDM. http://cdm.unfccc.int/Reference/Guidclarif/accr_handbook.pdf.*

Others:

Environmental Impact Assessment (EIA) Notification (2006), Ministry of Environment & Forest, Government of India.

Environmental Protection Act (EPA) (1986), Ministry of Environment & Forest, Government of India.

Environmental Protection Rules (EPR) (1986), Ministry of Environment & Forest, Government of India.

Guidelines on (i) Research and Development and (ii) Sustainable Development for Central Public Enterprises (CPSEs), (2011). Ministry of Heavy Industries and Public Enterprises, Government of India, pp. 5-18.

National Voluntary Guidelines (NVGs) on Social, Environmental & Economic Responsibilities of Business (2009), Ministry of Corporate Affairs, Government of India, pp.7-26.

The Air (Prevention and Control of Pollution) Act, (1981), Government of India.

The Forest (Conservation) Act, (1980), Government of India.

The Water (Prevention and Control of Pollution) Cess Act, (1977), Ministry of Law, Justice and Company Affairs, Government of India.